Birds of a feather

LESSONS FROM THE SEA

Kathleen P. O'Beirne
Lifescape Enterprises

Other Works by the Author:

Pass It On! How to Thrive in the Military Lifestyle
ISBN 1-879979-00-4

Pass It On, II: Living and Leaving the Military Lifestyle
ISBN 1-879979-00-4

Student Passport
ISBN 1-879979-06-3

Mobile Student Passport
ISBN 1-879979-07-1

Life is a Beach! Musings from the Sea
ISBN 1-879979-09-8

Published by:
Lifescape Enterprises
P. O. Box 218
West Mystic, CT 06388

Printed in the United States of America

10 9 8 7 6 5 4 3 2

First Edition

Graphic Design by Jeanne Sigel
Printed by The Day Printing
ISBN 1 –879979-02-0

To all who have struggled to understand themselves
and others, and to all who have found
peace and delight in observing
the shorebirds,
may this book bring you insights.

My great thanks to my patient and observant
husband, Frank (Mick) O'Beirne, Jr.,
who always finds the Green Heron first!

Kathleen P. O'Beirne

Prologue: The Seascape

But there are other beaches to explore.
There are more shells to find. This is only a beginning.

Anne Morrow Lindbergh, *Gift from the Sea*

Two of the most fragile shells on the beach,
To which I am drawn and readily reach,
Are the Angel's Wing and the Sailor's Ear:
Our ability to soar and our need to hear.

Kathleen O'Beirne

On a visit to Sanibel Island, Florida, more than ten years ago, I had the delightful experience of re-reading Anne Morrow Lindbergh's *Gift from the Sea* — the same copy in which I had underlined passages as a young woman in the early sixties. As a young wife and teacher, I had especially liked the sections on the movement from the early stages of marriage (the double-sunrise) to the oyster bed stage, as special interests, my career path, and volunteer commitments began to shape my life differently from that of my nuclear submariner husband. Her oyster analogy spoke to child-rearing years, and her plea for creative solitude would echo throughout my life as I tried to find her balances of solitude and connectedness with family, friends, and "causes" in which I deeply believed.

However, I found myself in a new place on the second reading, with children grown and away from home and no major responsibilities for my parents and parents-in-law....yet. My concern, in my early fifties, was how to thrive in corporate America. Women today are in environments in which they may be significantly more vulnerable than when most operated primarily within their own home and volunteered in their community. They encounter the world outside of their home in ways that their mothers and grandmothers did not, and often are ill-prepared to cope with behavior patterns to which their past experience has not exposed them. For those who knew the darker sides of human nature at home, as much as for those for whom home was safe and supportive, there is a need to widen perspectives and strategies for effective interface.

Recent demographics on women in the workplace indicate that the top fields (for non-military women) in order are: administrative support (almost one-fourth of working women); professional specialties; service work; executive, administrative and managerial; and sales. Those in administrative support often have a great deal of responsibility, but very little control (a bad combination). They often must interface not only with their boss, but also with the public. Those who are professionals and supervisors often have had very little leadership (vs. management) training and may not have experienced appropriate role models. The hunger for such information is palpable when one attends the enormously popular one-day seminars on "conflict-resolution" and "difficult people." Unfortunately, these one-day seminars provide only short-term effects, and all of the effort belongs to the attendee himself or herself, unless s/he has been fortunate enough to go with a small group from the office who can then support each other.

Deciding that knowledge is effective armor, I sought a new set of symbols from which we can learn in the same way that we learned from Anne Morrow Lindbergh's shells. I was struck by the fact that the islands of Sanibel and Captiva that had provided her with her wonderful insights, were the same that now would give me a gift for our time. The more I observed the birds of Sanibel, I became convinced of their similarity to people in our places of work, as well as our places of family, school, and volunteer commitments.

The birds are all concerned with gleaning their daily food, finding shelter, and maintaining their physical safety — the basic needs for all creatures, including human beings. According to Maslow's hierarchy, we cannot focus on the higher level needs, such as social and emotional relationships, intellectual, aesthetic or spiritual pursuits, until we are assured of survival levels of these basic needs.

Some of the differences among the fowl were seen in how they earned their daily bread. Some were frantic, others methodical, and still others keen observers. Some were generally solitary, while others were almost always in the company of others, either like themselves or, occasionally, mixed.

In the flocks, I noticed that some handicapped individuals could survive, though they were often on the slow end of the waders: the White Ibis with the one leg who had to hop instead of shuffling, and the Sandpiper with the injured foot who had to minimize his scurrying.

Some consume small amounts constantly, almost frenetically. Others wait and are selective. Some eat peaceably, knowing that each wave brings more food and it is sufficient. Some protect their space, fervently chasing off even the most innocent of intruders. Some squabble over every visible scrap, with bills clacking, wings outspread and threatening, and cries of alarm. Some feed independently side by side with other species; while others, such as the White Pelicans and Cormorants, have collaborative strategies.

As Anne Morrow Lindbergh noted in *Gift from the Sea* , an individual can grow from one kind of shell (relationship) to another, but not all do. She strained to achieve creative solitude in her own life; and while I believe that artists, poets, writers and, perhaps, self-employed business women do attain her ideal state with effort today, a far larger proportion of women seek meaningful achievement within corporate, academic, business, or other bureaucratic environments. Recognizing the variety of behavior patterns there and, perhaps, seeing ourselves in various stages of growth, may lead us to ponder appropriate strategies for shaping ourselves and for dealing with others. Understanding the modus operandi of each species will be enlightening. One may consciously choose to adopt some of the characteristics of a behavior pattern that is seen as desirable and modify some of the characteristics that seem less so. In this way, one may become a fairer fowl. However, each of us typically stays fairly true to an overall species pattern.

So, the idea of a guidebook to behavior patterns in the workplace was born. It would function much like the bird books I queried to be sure that my identifications were correct, as some of the species at Sanibel look very much alike in varying stages of maturity, different only in leg or bill coloring. Additionally, guidebooks help us become more expert over time and allow us to recognize individual idiosyncracies that make each creature different, even within the same species.

The environment in which shells or birds occur may affect a collection of types. For example, I've seen an area on the beach where all of the shell species had a strong yellow coloring not usual to their natures. They were unusual, but still recognizable as members of their species. In another case, there was a paprika-like algae all over the shells and sand, which made identification very difficult. So, too, generational, organizational and lifestyle colorations may impact members who work, volunteer, or worship in those settings. Sometimes the impact is beneficial; sometimes toxic.

I worried that serious birders might be offended by my attempt to anthropomorphize these birds, making them symbols of recognizable behavior patterns. In a meeting with Roger Tory Peterson and his wife Virginia, I asked if he believed birds had "personalities." He immediately launched into the personalities of some of his favorite birds, so I knew I was safe!

Sanibel Island is a remarkable place in which to observe the sea and shore birds in the wild. Not only do the J. N. "Ding" Darling National Wildlife Refuge and Tarpon Bay provide wild areas in which to encounter a wide variety of species, but the birds populate the public beaches as well. People have evidently treated their bird counterparts with courtesy and respect, because these birds show little or no fear of human beings. The Seagulls do not behave as they do at public beaches elsewhere, making pests of themselves begging and swooping for food. They follow their natural patterns. Likewise, a lone fisherman may find himself in the company of an almost man-sized Great Blue Heron, who watches his success with great interest and utters his thanks when treated to a fish.

Sanibel Island is also a place that allows one to stretch one's insights in peace. Symbols and metaphors seem appropriate for such a place. It is a place where poetry comes easily and where perspectives on thorny work situations benefit from distance and lack of distraction. While I will grant you that not all of us are given to introspection or sensitivity to others' behaviors, those who are will find this guide useful.

I will present the birds singly, including not only their physical appearance and habits, but their behaviors both fair and foul. Your own examples, observations, and individuals will create your own "Seescape." As you think about authority figures you have experienced, assess your parents and teachers in addition to leaders and peers in your workplace. A separate chapter provides a description of how a minister of each type might appear (which grew out of my experience on a pastoral search committee). Resumé, interview, and sermon styles are explored. These can be extrapolated to situations in which you must select new employees or new professionals, such as your doctor or dentist.

You will find a section on scenarios: how each bird perceives or behaves in a variety of situations. Then you will find an interface matrix that will give you a quick look at the interaction between each possible pair of birds with one as boss and the other as employee. Just as it takes you awhile to learn all of the shells, birds, or reptiles in a nature guidebook, and you must refer frequently in the beginning or when you have not seen a species for awhile, you will also learn these metaphorical birds by frequent sightings and practice.

I have taken fragments from a number of folks in each behavior pattern. Each portrait is a collage, so, if you think you decipher one specific individual, you really are noticing a cluster of behaviors that typify one bird. When a metaphor is powerful and accurate, many readers believe they see themselves or people they know in the descriptions.

While I dislike having to stumble over the grammatical issues of she and he (s/he) and his or her, we do encounter each fowl as an individual. I think you will identify the birds in your life, including yourself, more readily with this individual approach than with a plural composite.

Our tendency is to see certain of the types as gender-specific; but, they are not. Therefore, when they do not fit our stereotypes, they are even more puzzling to the average person who encounters them (e.g., the woman who behaves as a Stork or the man who is a White Ibis). That is why the use of the bird metaphor works so well — because most of these birds' sex is very difficult to determine, even for ornithologists.

Another way in which the bird guidebook concept helps us is that we focus on clearly identifying each bird we encounter. Symbolically, it helps us remember that people's behavior mostly has to do with them, not with us. Instead of rushing to blame ourselves and feel guilt over an interaction, a tendency of some of the metaphorical birds in this book, we must look at the others' feathers.

I recognize that some may choose to use what they learn here to manipulate others, to do ill. They will be relatively few. More will read because they are struggling to understand interactions with others that are not as fruitful as they would like. I hope that you will find strategies that enable you to be your best self, and also give you ways to work best with others who are different from you. Yet another lesson from the Sanibel beach is the incredible diversity of species found there, making it one of the richest shell and bird environments in the world. Businesses can learn from this model.

One of my favorite silkscreens by Ikki Matsumoto, who used to live and work on Sanibel Island, is of a frigate bird with swirls of current behind him. It is hard to determine if he is uplifted by the waves of air, or if he is barely keeping them under control behind? Is he on the leading edge or about to be overwhelmed? This dichotomy, rarely so beautifully defined, challenges us everyday in our interactions with others.

Many of my notes for this book were scribbled on church bulletins from this past decade. Clearly the challenge was on my mind and in my soul for years. Therefore, it is fitting to conclude:

> *Let us go forth into the world in peace, being of good courage, holding fast to that which is good, rendering to no one evil for evil, strengthening the faint-hearted, supporting the weak, helping the afflicted, honoring all persons, loving and serving the Lord, and rejoicing in the power of the Holy Spirit.*
>
> "Our Common Commission,"
> Mystic Congregational Church, Mystic, CT

> *One can never pay in gratitude; one can only pay 'in kind' somewhere else in life.*
>
> Anne Morrow Lindbergh, *Listen! the wind*

Kathleen P. O'Beirne
Mystic, CT

Contents

Tracks in the Sand

What was the sea whose tide swept through me there?
Out of my mind the golden ointment rained,
And my ears made the blowing hymns they heard.
I was myself the compass of that sea:

I was the world in which I walked, and what I saw
Or heard or felt came not but from myself;
And there I found myself more truly and more strange.

Wallace Stevens, "Tea at the Palaz of Hoon," *Poems*

With greater insights there is the prolonged wonder, not so much of being wise oneself as of having stumbled into a wise world. The mind feels suddenly integrated with the power of nature and moves at nature's pace.

Robert Grudin, *The Grace of Great Things: Creativity and Innovation*

There are tracks in the sand. On an observation trip, I came across human footprints occupied by a flock of Plovers. Each faced the same direction (into the wind) and looked very pleased to have a perfect resting place, however temporary it might be. The symbolism made me chuckle.

Another morning, I spotted a Great Blue Heron on the beach at Captiva. After moments of sheer awe, I watched him fly away. When I reached his footprints, they were as big as my hand spread out. One does not find tracks in the sand at the cutting edge of wave action. Only as the sand becomes increasingly dry can prints be made that last for awhile, until others traverse them or a high tide fills them in.

Where myth, metaphor, and methodical observation collide, we have a powerful tool for understanding ourselves and those around us. As with any creative endeavor, after the first flash of "aha!," there follows a period of observation and research. My observation was both ornithological and anthropological.

Notes were made on the birds at Sanibel Island over the course of a decade. Species' behaviors were verified with a Fish and Wildlife Service ranger at the J. N. "Ding" Darling National Wildlife

Refuge and through books on birds. But, really very little has been written in this field, so persistent personal observation was critical. The human parallels were observed in tandem, often with great delight, and sometimes with great puzzlement.

The research demonstrated that the desire to type people by behavior categories is as old as mythology and as new as the best seller list. Throughout history, we have sought ways to exemplify, personify, and clarify the range of behavior patterns experienced in our daily lives. One finds very elaborate archetypes in the Greek and Roman myths. These archetypes were taught to each generation through art, drama, and literature as part of their conscious cultural heritage. Hippocrates sorted people into four categories based on their predominant body humors. The Sufi Indian tradition of personalities has been adapted by modern thinkers in the form of the Enneagram. The Lakota tribes, of the area now known as North and South Dakota, had their own four-part wheel that depicted the personality of each season with an animal or a bird.

Carl Jung, who wrote about our "collective unconsciousness" derived from myths and stories, identified cameo personalities. His work was codified by Isabel Myers and Katheryn Briggs in their Myers-Briggs Type Indicator; and David Kiersey and Marilyn Bates further modified their work with four major temperament types captured through the Kiersey Temperament Sorter.

In the Christian tradition, Jesus and his twelve disciples can be viewed as having distinct and disparate characteristics. The various stories of the Knights of the Roundtable treat us to a range of personalities. Even children's literature has adult levels of interpretation that demonstrate patterns of behavior. We see a rich array of types in Winnie the Pooh and his friends, Alice and the creatures she encounters in Wonderland, the characters in *The Wind in the Willows,* and the fairy tales and fables. Once you become comfortable with the Sanibel bird collection, you might test your self by drawing the parallels with some of these familiar stories that you probably absorbed at a different level the last time you read them. Then, keep an eye on cartoon series such as *Peanuts* and *Dilbert.*

Forensic handwriting experts use a typology. One of the most-read sections of the newspaper is the daily horoscope section. The best seller lists have been full of books for the last decade designed to help readers sort out themselves and others by colors, interior decoration preferences, eight American types to help marketers target their products, the left-brain/right-brain/integrated-brain trilogy, birth order and its impact in the workplace, and family roles that people carry over into their other activities. These schematics have run the gamut from the too-simple-to-be-used accurately to the too-complex-to-be-used easily.

Nancy K. Austin wrote in an article called "Managing by Parable," "As every psych major knows, human beings reason largely by means of stories, not mounds of data. Ten years ago, in writing *A Passion for Excellence,* Tom Peters and I discovered that top-notch leaders think of themselves as master storytellers: they use anecdotes to teach, chart complex social territory, perpetuate important values and point the way through change." She said that the Stanford Business School urges its students to study the Enneagram types in order to understand employees, customers, and organizations.

Many of the typologies have been rather abstract or clinical in nature, and the average individual has difficulty storing that information in a way that can be accessed readily. When the typology has a cohesive visual component, such as the birds at Sanibel, it is easier to remember and to use when confronted with a complex or volatile situation. Americans have a passion for bird-watching. It is the second most-favored hobby in this country, second only to gardening.

CAUTIONS

As one begins to use a typology, there are several warnings. One must be careful not to confuse a *stage* of growth with the more permanent *style* of living. Also, one needs to be wary of labeling children and interacting with them based on our assessment, because we may indeed shape them accordingly. Thirdly, one needs to be alert to the dangers in calling certain behavior patterns male or female. And, finally, do not be tempted to assign higher or lower intelligence (IQ) to given birds, though it will certainly be appropriate to find a higher or lower emotional intelligence (EQ) among the species.

In the late 1970s, research was done on the four learning stages through which children progress. Intriguingly enough, these stages also are seen in adult groups. They basically move from the first chaotic, kindergarten stage of interaction with everyone in view to the second stage of more focused interaction with the primary authority (be that parent, teacher, or boss). This is a stage of rules and fairly civilized, non-critical behavior.

The third stage, well known to all parents of pre-teenagers and adolescents, is that of pseudo-independence. Ironically, the rebellion is highly dependent upon the known values of the authority figure (even though the choice may be 180 degrees out). Because of some inappropriate behavior in the third stage, there is a temptation to throw out the baby with the bath water in stage three and preclude a healthy progression to stage four.

The fourth stage requires the emergence of a cognitively independent individual who is aware of the values and expertise of others; but, in the final analysis, makes a synthesis that is truly his or her own. This rare and wonderful creature is the stage for whom this democracy was built. Our Founding Fathers would not have stated it in these terms; but then again, they are a collection worthy of our analysis! As Anne Morrow Lindbergh noted in her *Gift from the Sea,* not all people reach her ideal nautilus shell state.

Children are creatures-under-construction. If they hear labels early on, they may find it easy or wise to comply. There probably is no more sculptural, Pygmalion-like setting than our own family. When we hear over and over that we are not the generous one, we may fulfill that prophesy (certainly in relationship to those who delivered this interpretation), even though our inclination might be otherwise. Children observe others' body language and tone of voice which comprise 87 - 93 percent of human communication.

It is important that we allow people to choose their style (their preferred behavior pattern), regardless of their gender. Does this make it harder to identify them? Yes. Why? Because we are coming out of a period of time that wants to pigeon-hole folks (another bird analogy!).

However, some researchers, such as Daniel Goleman, do find evidence that women's early socialization may lead to their being more sensitive psychologically and physiologically to the behavior and comments of others than their male counterparts. If this is so, their sensitivity packs a double-whammy in the reading of others and the acceptance of others' assessments. If women are consistently more accurate than men at interpreting unspoken messages in gestures, facial expressions, and tone of voice, then they show greater skill in figuring out how the people they observe are feeling. While men are more likely not to notice others' feelings and assessments of themselves to begin with, and to slough off any negative they may perceive, women, especially adolescent girls, are more likely to ruminate over hurts, slurs, and slights and are more prone than their male counterparts to depression.

Actual raw intelligence is likely to be fairly evenly spread among the various behavior patterns. However, it is tempting to see the Green Heron, who is the quietly analytical bird, as more intelligent than the Brown Pelican, who seems to bounce happily through life without a care in the world. The Snowy Egret may believe himself or herself to know more than others and to be more precise in organizing or documenting what s/he knows, but that may be a very small universe. An individual's outward style should not be confused with cognitive ability.

A given individual may migrate from one behavior pattern to another in abnormal circumstances. For example, in periods of great stress, or conversely, of great joy, there may be situational change. The key is to respond as appropriate for the pattern being demonstrated.

IF THE FEATHERS FIT, WEAR THEM: REASONS TO UNDERSTAND BEHAVIOR PATTERNS

As Abraham H. Maslow wrote in *The Farther Reaches of Human Nature*, "We must remember that knowledge of one's own deep nature is also simultaneously knowledge of human nature in general."

At a very personal level, it is important to understand yourself. It is reassuring to know that while you are indeed one-of-a-kind, you do bear a striking resemblance to many others in this society. As you understand yourself and others better, you may be more adept at surrounding yourself with those who are good for your mental health, and removing yourself from those who are not, be they in your workplace, your volunteer place, or your family. Research is showing us loud and clear that working for or with someone for a long period of time to whom you react stressfully results in a lowering of your auto-immune system (i.e., you get sick!). More subtle are the energy sappers; these vary by your own pattern, but the result is that they wear you down or out. Therefore, it behooves you to learn how to interact strategically with other people; and, part of that is knowing when to leave.

You may also learn to recognize your own "buttons" or "land mines," so that you can see what role you play in any interaction with others. This will help you avoid setting the stage for negative results.

Each of us has a tendency to believe that others are really like us deep down. If they were less stressed, of our same generation or gender, or came from the same socio-economic background, they would really be the same. This is simply not true. The motivations, fears, and values of others are usually quite different from ours, regardless of stress, generation, gender, etc. There are those who really are very uncomfortable with any kind of confrontation; while others simply have to get their daily supply of confrontation by noon to get their adrenalin going, or they will prowl or probe until they get it.

WHY STUDY BEHAVIOR PATTERNS?

If you are in a position to supervise others or to hire and fire others, you may gain insights on how to select the best individual for a given position or how to infill in an already existing team or organization. You may learn how to get the best performance by happy workers because you have learned to assign tasks and interact with them in ways that are most appropriate to their particular style of behavior. Instead of following the old Gilbert and Sullivan adage, "Make the punishment fit the crime," you can make the reward or promotion fit the gifts. What a radical and effective concept! It is usually the most efficient as well, certainly after everyone gets the swing of things. How often, for example, has our society promoted teachers who love their classroom, doctors who are especially effective with their patients, social workers who thrive on their interaction with clients, and an inventor who has made quantum leaps in his field to the next level of seniority: administration? These folks begin to burn out in their new position, doing tasks that they do not love and at which they are not particularly gifted; and the organization begins to wonder how they could have goofed in making such a choice.

Additionally, as our workplaces become more diverse in terms of ethnic, cultural, educational and religious backgrounds, we will still have the diversity of behavior patterns to deal with — only now with layers of camouflage that we have barely begun to decipher. It is important not to determine that a given pattern is driven purely by the other forms of diversity, thus adding an inaccurate stereotype. Folks who are comfortable with change will be key players in the organizations of the future.

RAGE AND RUDENESS IN THE WORKPLACE

Many organizations are dealing with rage in the workplace or, at the least, uncomfortable and inappropriate confrontation among co-workers. Part of this may be that an individual is still in stage three and has not learned to separate cognitive critical thinking from critical behavior. Part of it may be that certain behavior patterns rub salt in wounds. But, the high tech/high touch combination has gone sour in many settings, with the touch part being the straw that is breaking the back of mental health and productivity . One does not need to look very far for the source. Dean and Mary Tjosvold, in *Psychology for Leaders: using Motivation, Conflict, and Power to Manage More Effectively,* report that 60 - 75 % of employees say that their immediate supervisor is their greatest source of stress, and "three-fourths of the highly successful executives in three *Fortune 100* corporations reported that they had at least one intolerable boss during their careers." The Gallup Organization, after polling employees worldwide for 25 years, reported that people don't leave companies, they leave managers.

CUMULATIVE IMPACT

At issue here is the cumulative impact of such experiences. If one has back-to-back intolerable experiences, the impact can be extremely damaging. General Electric CEO, Jack Welch, in his final letter to shareholders, wrote that companies should "love and nurture the top 20% of their employees, but actively weed out the bottom 10%." He went on to say that the top 20% should be "rewarded in the soul and wallet because they are the ones who make magic happen. Losing one of these people must be held up as a leadership sin."

However, because of the legalistic, litigious nature of so much of our workplace interface, it is very difficult to remove bad actors whose performance by position measures is at least mediocre. Often those who have the most problematic attitudes and behaviors are those who announce to one and all that their attitude does not negatively impact their customer service or production. This is usually very hard to prove concretely and is issued as a dare or threat. These same folks usually have encyclopedic knowledge of all of the union, equal employment opportunity, and human resources department regulations.

The truth is that they do negatively impact their co-workers; but, as this is also hard to prove concretely, it is critical to: 1) recruit carefully; 2) retain them only if they are truly working out well in all domains when the organization's no-fault release point is reached; and 3) recognize and reward only the outstanding employees. Less-than-outstanding employees need to see diminishing returns on their horizons unless they turn themselves around.

The pendulum has swung too far on work employment protection, making it difficult for managers to protect their "stars" from bad apples. Current policies and procedures almost guarantee that those with talent and commitment to excellence will get up and go, leaving an organization with the least ept and the mediocre. The bad actors usually have a canny ability to recognize the noose at the end of the loose rope you are giving them to hang themselves by. Therefore, we must get better at assessing individuals and their probable impact when we are choosing new employees, or when we ourselves are choosing new employment settings.

FAIR AND FOUL

You will read about each fowl's behaviors, whether fair or foul. The same pattern of behaviors that can be positive and successful can, at the other extreme, run amok. (For our purposes, it might also be amuck!) Only when those behaviors are conscious, purposeful, and malicious does the fowl lose his or her original designator and become a Cormorant. The Cormorant is the next-to-last bird to be discussed in this book because s/he is the evil version of the other patterns. Many of the traditional typologies do not have an evil category, choosing instead to keep a gravely dysfunctional person within his or her type. However, once folks slide into the malicious mode, they adopt specific additional behaviors above and beyond their original pattern, so need to be categorized separately.

While the issue of good and evil is as old as our myths, religions, Dante's *Inferno* and Milton's *Paradise Lost,* many traditions recognize the existence of those whose focus is doing harm to others. Languages other than English have verbs that make the distinctions clear. For example, in Spanish, if one says "Está mala," it means that s/he is behaving badly today. But, if one says, "Es mala," it means that s/he is evil, not just today, but every day.

As you make the decision about whether an individual has become a Cormorant, you may struggle over the conscious vs. unconscious acts. You may tend to forgive one or two conscious acts; but, at some point, for your own well-being, you will know at a deeply emotional level that you have a Cormorant on your hands. Having documentation of the cumulative behaviors is necessary as well. For example, it can be difficult at first to assess if the behavior you have experienced has been "just" the arrogance of the Stork or the cold precision of the Snowy Egret at work, or the more hurtful malice of the Cormorant. While the two former can be unpleasant to work for or with situationally, the Cormorant is a destructive force in the workplace and is to be avoided or constrained, depending on your level of authority.

Once a Cormorant, always a Cormorant? Perhaps not, depending on his willingness to change his or her ways and admit, at least to himself or herself, that s/he has been a Cormorant. However, you would be wise to follow the strategies outlined in the Cormorant chapter to protect yourself from this very wily and usually unrepenting bird.

A CHANGE OF FEATHERS

Another phenomenon of which to be aware is that while people do tend to operate from one pattern most of the time, a change in circumstances, such as a company retreat, may encourage what seems like a migration to another profile. It probably is just to the fair end of his or her spectrum. If, for example, there is less stress in a retreat or social situation, someone who is aggressive or reserved at work may be more playful or more companionate. Typically, those who behave situationally will revert when back in the usual environment in which you encounter them. So, be wary. Their memories may last longer than their kindnesses, and their predator tendencies may make you vulnerable if confidences were shared in the less stressful mode.

Psychologists differentiate between *traits* or *dispositions* (our enduring characteristics) and *states* (which are temporary conditions). When people proudly announce that they are completely different in settings away from work, beware. Who is the real person? Who is the facade? or charade? There is manipulation here. Which of the birds demonstrates this characteristic? This is different from flexibility. Their behavior is aberrant, unstable, or lacking in continuity at best. The bottom line is that you need to engage with these folks as they behave where you encounter them. That is reality for you.

On a more positive note, folks under stress in their workplace may not feel able to exercise their playfulness. This, too, is a warning sign, especially to the individual himself or herself, and to the supervisor. Whenever there is a disconnect between the way an individual behaves in different settings, there is dysfunction.

A PARADIGM SHIFT

The nature of heroes varies by time, place, and circumstance. The behavior patterns that may have been the most fitting for given periods of history may no longer be the most valued today. For example, after hearing the description of the different temperaments found in the Myers-Briggs typology, and especially after hearing that general officers and admirals tend to have an "N" in their four letter code, one of my senior officer students at the National Defense University queried how he could get to be an "N?" While he was both amusing and rather pathetic at the same time, I did point out that he would not become an "N." However, he could mimic or pursue the behaviors that he admired, and with practice-becoming-habit, he could appear "N-like." You can make choices and act positively on them (c.f., Glasser's Choice Theory).

By understanding how people behave, you can gain a perspective that puts distance and, hopefully, wisdom between the immediate, puzzling personal interactions you have and the potential positive interface you could have. You may gain patience and perseverance to improve relationships. And, finally, you will gain power for positive, pro-active strategies.

Hopefully, the behavior patterns you find ahead, plus the scenarios and interface matrix, will leave you with an "aha!" Understanding the birds of Sanibel will aid you not only in your workplace relationships, but in courtship, in your relationship with your parents and in-laws, and your interface with your elderly parents or grandparents (who become more polarized for good or ill as they age). When you search for new pastors, counselors, dentists, doctors, or your child's piano teacher, do not select them purely on the basis of their expertise, because your ability to consume their expertise depends upon your positive reaction to their behavior patterns.

HOW TO USE THIS BOOK

First of all, seek to find yourself among the assorted patterns. Intra-personal intelligence is one of Howard Gardner's eight intelligences. Self-knowledge is not common to all of the patterns, so you may find this harder than you expect. Knowing your positive attributes, as well as the ways that you negatively impact yourself and others, enables you to choose how you behave.

Then, inter-personal intelligence, the understanding of others, can help you choose how to behave with others. Again, making this choice is not common to all of the profiles. Some have very little interest in their affect on others. However, such a stance does not augur well for one's success in today's workplace, community, or family.

Perhaps a useful way to approach the chapters ahead will be to read one a day and take time for assessment and absorption. Then spend the next day looking for folks who fit the profile, practicing your strategies with them. Add your own marginalia, because this should be a working reference to which you return as you decipher folks you already know and those you newly encounter. See the space at the end of each chapter to record your bird list.

However, there will be those who by the very nature of their own behavior pattern will not find this information interesting, valid, or useful. They bring to mind an elderly couple entering the dike area at the J. N. "Ding" Darling National Wildlife Refuge at Sanibel Island. My husband and I had just been struggling with whether we had seen an immature Yellow-crowned or Black-crowned Night Heron. The wife said to her husband, "Are there any penguins here?" Not everyone will understand or care. But, these birds may make a world of difference for you.

Wood Stork

Wood Stork

The Wood Stork, the only American Stork, is relatively common in southern ponds, lagoons, swamps, and marshes. One of the largest of the Sanibel Island birds at 34 - 47" tall, he has a very large wing spread of five and a half feet. He is beautiful in the air, alternately flapping and gliding with his neck and legs extended; his wing beats are slow and powerful. When travelling great distances, he often soars very high on the thermal air currents. His glide without flapping resembles the glide of an eagle or vulture.

He is mostly white, with black on the wings and tail. His dark head is naked, very much like that of a vulture (his closest relative), and his bill is long, thick, and curved downward. The Flamingo also belongs to the Stork family.

When feeding, he strides with his head down; or, he may run about, scaring off all other bird species from his chosen feeding area. Unlike some other species, the Stork rarely demonstrates patience. He sticks his bill into the mud and then muddies the area with a reddish foot, literally stirring up the muck. He sweeps his bill methodically through muddy water or mud flats. When his sensitive bill locates food, it snaps shut with one of the fastest reflexes among vertebrates because his prey is fast fish vs. the worms and mollusks that other probers feed upon. His white under side is a helpful camouflage because his prey cannot see white against the sky as easily as a darker color. Sometimes, several Storks together form a phalanx and flush fish from shallow ponds methodically.

His primary foods are fish, reptiles, and amphibians. His diet includes everything from grasshoppers and mice in dry grassy areas to a baby alligator, fish, frogs, large insects, snails, and sometimes plants in moist areas. He will also scavenge dead fish and carrion. He can feed in murky waters (vs. the Herons and Egrets who watch for their prey). Storks feed best in small bodies of water where fish are concentrated because the water is drying up... sounds like corporate mergers!

Although ornithologists indicate that he is usually silent as an adult, as an adolescent he has a raucous hoarse croak and can be heard even before he is seen, when he is squabbling over food. If he is with others of his own species or those of other species, he wants what is between their toes and is very aggressive about getting it. He will even use his bill to strike an offending competitor on the head.

FOWL FOLKLORE

The Wood Stork was sacred to ancient Egyptians. Because his long, curved bill looked like a crescent moon, he denoted the moon god, Thoth, the god of wisdom and learning. When an Egyptian ruler was buried, a Wood Stork was often mummified and entombed with him. Part of the mythology is that Thoth became a Wood Stork in order to escape the fire-breathing monster Typhon. Poseidon, the Greek god of the sea, sent Thoth to Tenos to clear the Aegean island of snakes. The Stork is one of the few human behavior patterns who can take on the Cormorant (the snake bird) and win, so the mythology works on several levels.

Other Storks in mythology are Athena, the goddess of wisdom, and Zeus, her father, the supreme ruler of the gods. Zeus, known to the Greeks as Jupiter, was Lord of the Sky and used the thunderbolt in his roles as Rain-god and Cloud-gatherer His symbol was the eagle, whose niche the Stork occupies in this study of behavior patterns. Zeus himself gave birth to his daughter, an interesting parable for daughters of powerful men throughout history. Zeus was also known for his philandering and his attempts to hide this from his wife, Hera.

For the Lakota Indians, the stork niche is filled by the buffalo, a strong leader who could be stubborn or arrogant.

The Stork that we commonly associate with the delivery of babies, is actually a different creature native to the Mediterranean.

OBSERVED FOWL BEHAVIORS

Wood Storks are indiscriminate bullies. They operate on a “might equals right” philosophy. They muscle in on smaller

species and do not even honor their peers' "space," often poking their bills right between the feet of a fellow flock member, hence enraging each other continuously. In human behavior, they call this keeping others "on their toes." There is much bickering, hissing, crashing and flapping of wings. Like "good old boys," they enjoy being with other Storks and measuring themselves in their effectiveness. They feed on what the others have stirred up. But, they do not like having the smaller species around and will actually give them a hard poke on the head.

Storks, unlike some of the other large species such as Herons, fly with their neck extended and fly great distances daily, scanning for food sites (often determined from the air based on the presence of other smaller species).

OBSERVED HUMAN STORK BEHAVIORS

For much of the history of this country, Storks have been the brave explorers, warriors, corporate giants, and leaders in many domains. Their confidence, competence, and commitment to mission have made them mythological in stature. They can spring into immediate action and have little difficulty in making decisions.

General George S. Patton was such a leader. Though described by Lance Morrow in *Smithsonian* as a "martial peacock" who "filled the stage with strut and plumage and flame and conceived that battle was essentially a dramatically amplified projection" of himself, he was brave and bigger than life. "He thought that he had fought with Alexander the Great in another life."

John Adams, whose wonderful biography has become a must-read for many Americans, was a Stork who set aside his own personal livelihood and legal profession to devote his prime years to shaping our new nation. Thanks to David McCullough's profile, we know the personal struggles that this deeply thoughtful man experienced, frequently at great distance from his wife Abigail, truly his soul mate. The sacrifice that they both made for this country have been little understood or appreciated until now. His sheer determination to keep the movement for independence on a noble plane makes clear his understanding of the magnitude and the symbolism of the task. He was blunt in speech, pragmatic in

outlook, tough physically, could argue legal issues with great clarity, regardless of the rectitude of his client, and enjoyed the company of the giants of his day.

He was the perfect counterpart for Thomas Jefferson (a Great Blue Heron). Together, they provided the whole that neither could have produced separately. When stressed, Adams studied his Greek, Roman, and French antecedents, much like a Green Heron, with scholarly tenacity and perseverance. And, Abigail, a remarkable Seagull, kept him balanced and fiscally grounded. His desire to be valued became especially strong after his retirement. He wanted to set the record straight. Though willing to own mistakes that he had made, he wanted to rectify erroneous accusations. He was particularly mindful of history's view of him.

Often these heroes have performed almost solo, with perhaps only a trusted companion or aide. They have been tough birds to live with and work for, with their predilection for command and control. Anne Morrow Lindbergh wrote of her husband's "terrific drive which he applies without discrimination to crossing the Atlantic, writing a book....or finding out the price of butter." His motivation was conquest. He expected attention to be focused on him. The hero's sheer energy and animal strength and magnetism made him noble, courageous, very exciting and also very beautiful. "But," she said, "it is not everything." A friend commented in A. Scott Berg's biography of Lindbergh that he did "not understand the light approach to anything... Perpetual emotion helps keep one from examining one's feelings."

Storks have traditionally fared well when they have assignments of short duration before moving on to the next higher level of responsibility. In the two to three year assignment, the Stork gets down to business as soon as possible, producing the product or service as quickly as possible. Because of the known short time frame, the Stork is relieved from the responsibility of building long-term relationships with people or building long-term processes for improvement. S/he simply needs to do well enough to get his or her ticket punched. Those who work for Storks in these settings often have longer periods of commitment and learn, sometimes painfully, to simply wait out a Stork whose behavior is

particularly foul. If senior leaders are not perceptive, they will find a hollow force when they most need core strength.

The Human Stork is often of significant height and strength, so s/he rules the workplace with his or her sheer physical presence and high energy. Sometimes, the Stork is more of a Bantam Rooster in size, but the Napoleonic complex makes up in perceived personal power whatever is lost by his or her diminutive height. The vocal prowess is often louder than necessary for the current task, is more punctuated in its pattern of delivery, and often is accompanied by accusatory, poking-like gestures.

One Stork affectation is a very abrupt response to others with the intent of demonstrating, "I've heard you!" and "Don't insult my intelligence." It is also an indicator of disinterest in details. "Just give me the big picture." Interruptions of others' discussion are constant. Eye contact is often used to intimidate others. And a knee-jerk response of "I disagree" may be used frequently as a mechanism for putting others on the defensive while the Stork has a chance to sort things out. Facts and analysis are not considered sufficiently "big picture" for the Stork, and the "bean-counter" is an anathema.

A Stork can be a leader admired from afar, often with almost charismatic charm, but s/he evokes powerful negative reactions among those with whom s/he interacts regularly. Autocratic, s/he is used to getting his or her way. S/he will blithely make promises, knowing that s/he either cannot or will not make good on them.

There is rarely any doubt or public dissension about what s/he wants. If s/he is not successful in obtaining the desired results immediately, s/he may flap his or her feathers symbolically or swagger off, waiting to combat at a more propitious moment. A senior Stork will keep subordinates, including other Storks, on a very short leash so that all power and glory go to the appropriate place. If one cannot afford to be tarnished, one has to polish the mirror and look at one's reflection with frequency.

On the other hand, a Stork who feels particularly comfortable with those around him or her, often is physically demonstrative in his interaction. For example, the big bear hug or closed-fist punch

on the upper arm are favorite signs of affection. Today's workplace prohibitions on touching not withstanding, the Stork likes to make contact with friends that way.

The Stork is a status leader who has gained power through a hierarchy. S/he is not usually chosen as an emergent, natural leader except in situations that call for immediate control and action, such as organizational transition, chaos, or war. S/he dislikes working with democratic or collegial committees, and especially dislikes strategic planning and "total quality management," which require long-term planning and processes. Short-term, immediate action is much preferred. S/he sees himself or herself as a realist and is usually opportunistic, as opposed to being creative or innovative. The Stork can be very provoking, as opposed to thought-provoking; s/he does this for purposes of power, not for the love of ideas and the democratic process.

In a Duke University Medical Center study reported by Tim Friend in *USA Today* in 1997, "Dominant men — the ones who interrupt conversations and feel compelled to be the center of attention — die younger from heart disease than men who are more thoughtful and laid back.....they are driven by insecurity to be on top. They're often considered obnoxious and not well liked by their peers." The researcher, psychologist Michael Babyak, said that one possible reason for the early deaths is that "they stay juiced on their own adrenaline, keeping their arteries bathed in stress hormones." The study did not include dominant women, but they are perceived to be at less risk because they are likely to be more collaborative and less competitive than their male counterparts. For women whose behavior pattern mimics the male Stork, this is a cautionary tale.

A Stork sometimes aspires to greater formal education or intelligence than s/he has. One clue to this is the frequent use of malapropisms, words that s/he may have read but neither knows how to pronounce nor use correctly. When introducing others whom the Stork sees as less powerful than himself or herself, insufficient care has been taken to learn how they pronounce their names. It really does not matter to a Stork. If allotted a specific amount of time for a presentation, a Stork will take as long as s/he

chooses, the schedule be damned. The Stork is both determined and dominating.

A Stork does not dissemble; s/he just takes over. The Stork is usually an extrovert and is sure that people will be interested in what s/he has to say. S/he gains energy from being in charge. S/he sees things as black or white and operates on a win-lose proposition. Let there be no doubt; the Stork must win. S/he demands Respect with a capital R, while most others are willing to grant or expect lower-case respect until the higher level is earned. If a Stork is experiencing negative pressure from further up the hierarchy, s/he will be quick to shift that pressure or blame laterally or down the food chain.

Rules are for someone else. Storks make rules for others. Explanations of his or her decisions or consensus-building are not part of his or her preferred behavior pattern. Micro-management through rules and the new computer technology has become a favored venue for Storks and for other bird profiles. This style often demeans the more able members of a Stork's staff; coupled with the perspective that one inspects what one expects, the Stork believes that s/he has indeed produced all of the positive work, so there is no need to reward others.

A Stork can be very volatile in moods, making shoot-from-the-hip decisions. There is no doubt when s/he is angry. A Stork distances himself or herself from other species, abiding their presence only as long as they do not interrupt or challenge his or her stance or preoccupation. Observers of human storks tell of having been summoned to talk to a Stork while s/he is working on the computer and being totally ignored. This can happen with some of the other birds as well. However, the rudeness that accompanies this behavior, with an abrupt dismissal after the briefest conversation, will usually identify a Stork. Other Storks simply get up and leave their own office; those left behind know where they stand. Some Storks do not feel obligated to observe common courtesies.

Fear of failure can be a major motivator for a Stork. That fear can be so central to a Stork's behavior that it drives him/her to

succeed in entrepreneurial domains that amaze others who are more cautious. This means that the Stork is always scanning the competition and the horizon. S/he is ready to act if opportunity presents itself and will not be deterred by details, legalities, or competitors. Risk-taking is part of the persona. The bigger the risk, the better.

A Stork wants what you have because you have it. There is a sense of scarcity. S/he is always protecting himself or herself, walking around like a tank, well-armored against whatever might come his or her way. A Stork is both defensive and offensive. S/he surrounds himself or herself with lesser, non-threatening folks, whose rank clearly establishes their diminished stature, while courting powerful connections, especially those with whom s/he cannot afford to be an enemy.

Although ego-centric, a Stork is usually is not given to introspection. Nor is s/he sensitive to others. S/he rarely knows what s/he is really feeling or why, and does not care. His or her variety of feelings is rather limited, but the experience of anger or delight is huge. S/he misses his or her own nuances and those of others. If s/he does take an interest in others, it is usually for information that s/he might find useful, as opposed to being truly interested in others' concerns. Those who love Storks and/or work for them must be very precise in making their own wishes and needs known. They must not wait for a Stork to notice.

Those who have observed human Storks in courtship say that they are very much at risk because they are flooded with feelings they do not understand. They want to get this relationship settled, quickly, with great gifts and a fancy wedding, if necessary; but, then let them get back to normal as soon as possible. The birth of their first child may evoke similar emotions.

Storks distrust those who do have a sensitivity to themselves and others and seek flexibility in policies and positions. They are seen as less committed to the mission, lightweights, Pollyanas, touchy-feely types, manipulative, and even untrustworthy. Storks call these folks "Teflon men." Storks use name-calling as a weapon.

A Stork believes that people with talent (i.e., junior staff talent) are simply commodities that can be found whenever necessary. Replacements for losses will be easy. The only time a Stork is impacted by loss of staff is when s/he thinks a competitor stole his or her property. Then, comments will be made about the deserter's loyalty, although the Stork may feel no loyalty.

Hurts, slights, or put-downs are dealt with in three ways: they are totally ignored or brushed off; they are allowed to ricochet off onto others; or, they are ammunition for an immediate, intense counter-action to get even. A favorite saying by Storks, after dishing out unpleasant news or receiving it, is "Get over it." Because the Stork uses confrontation as a tool of choice, s/he does not take its impact as seriously as other bird types who dislike confrontation and will do almost anything to avoid it. Guilt is not a frequent experience, though there may be a brief sense of having goofed and having been vulnerable.

Some female Storks bought into a belief some years ago that one had to keep a constant scowl or sneer on her face to convince others of her competence and to keep others at bay. Smiles were considered simpering signs of incompetence or insincerity.

A Stork has a very hard time relaxing, because s/he works hard and plays hard. Competition is at the root of everything. Games are a source of delight, but the Stork must win.

VISIBLE SIGNS

A Stork's office will tell you a great deal about him or her. There will be pictures of the Stork with other well-known Storks, plus awards and the equivalent of battle flags, depending on the organization. The desk will be large and facing the door. The furniture will relay status. A Stork likes outward lavish displays of wealth, achievements, and, perhaps, a "trophy spouse" or family. While the trophy female spouse has been well known in corporate circles, especially for the second marriage, one sees the phenomenon now among female Storks as well. They will choose particularly attractive partners, sometimes younger, whose major contribution to the duo is their consort duties.....a good travelling companion who mixes well in corporate social circles and,

perhaps, provides a father for the children that the Stork wants, but hasn't got time to nurture or enjoy.

A Stork likes uniforms, be they military or corporate. S/he likes to be able to read another's status, whether by stripes on the sleeve, insignia on the collar, medals on the chest, name tag (or lack thereof), or power tie and suit. Because a Stork does not read others well, s/he likes to see outward and visible hierarchical signs. S/he also invests in status symbols so that others will not be confused about her achievement. These can be items like a watch or other significant jewelry and a car — everything from a Porsche to a Humvee. A Stork who hunts, golfs, or rides makes sure that symbols of these avocations are visible in his or her choice of vehicle, clothing, and office decor.

HUMOR

Humor is both a source of entertainment and a weapon for a Stork. Jokes are collected and polished so that s/he always has a ready supply. Rarely spontaneous like some of the other birds, s/he is methodical in using jokes to open presentations or to entertain socially. A Stork's sense of humor can be sarcastic, caustic, demeaning, and often not politically correct. Bull-in-a-china-closet is a fair assessment. Nuances are not part of his or her vocabulary. A Stork's interpersonal skills may be summed up in one-upsmanship. S/he is stubborn, self-focused, insensitive, but usually not malicious. A classic statement by a Stork is, "If you can't dazzle 'em with brilliance, then baffle 'em with bullshit."

POLITICS

A Stork believes that there is only one way to vote. The other party is of no consequence. Often part of the good old boy/girl network, s/he feels annointed in the position held/sought and, therefore, is vulnerable to competitors' solid analysis and charismatic leadership. Not deeply democratic, this office seeker/holder can be arrogant and out of touch with his or her constituency. On the other hand, because so many other powers-that-be are of the same ilk, s/he has a potent brotherhood and can, practically speaking, achieve benefits for the home district on a *quid pro quo* basis.

In social situations, a Stork presumes that others are of the same political sway as s/he, and can be obtuse in his or her lack of sensitivity for other points of view.

Stork politicians sometimes have manipulative advisors or spouses who are getting their power symbiotically. Because the Stork is so attached to this advisor, especially if s/he has entered a new arena of activity and is not yet sure of the lay of the land, it takes a great deal of skill to separate what the Stork really wants from that of the grooming fish. Similarly, a woman who finds herself senior to her husband in the workplace, may be treated to spousal advice such as "truly efficient leaders have no papers on their desk" or "truly effective leaders delegate all responsibilities to free them to be responsive to their boss."

FAIR FOWL

The positive end of a Stork's continuum gives us the corporate leader who has realized that the old John Wayne style of leadership will not be an effective way to grow successors. The fair Stork has moved from management (doing things right) to leadership (doing the right things) and has learned some positive inter-personal skills that enable those around to contribute in a meaningful way without fear. Those of different species are tolerated, and even included at the table.

Theodore Roosevelt, a classic Stork in his young days if there ever was one, became a fair fowl as he aged. His advice: "The best executive is the one who has sense enough to pick good men (sic) to do what he wants done, and self-restraint enough to keep from meddling with them while they do it."

A fairly recent development is the emergence of the senior corporate officer whistle-blower who recognizes that his or her company could "do it right" and not play games with accounting practices. The motivation is that the company to which s/he as committed his or her energies and faith is being undermined by ignoble leaders.

Those who seek to take a fair female Stork down a peg or two, will refer to her as "ambitious," as if this were a new scarlet letter

of some sort. The motivation of those seeking to denigrate is jealousy.

FOUL FOWL

Storks are sometimes described as being charming, brilliant, and even visionary. This can be camouflage for the ruthless, petty, careless and cunning behavior attributed to the same individual. High profile positions seem to attract Storks who behave much like the little girl who had a little curl in the nursery rhyme: "When she was good, she was very, very good; but, when she was bad, she was horrid."

Observation of members of Congress can baffle us. How can the same person run the gamut of the noble to the profane? One member of Congress, referred to by some as a Dragon Lady, left a senior colleague sputtering: "I have been told about your charm and wit... The reports of your charm are overstated, and the reports on your wit are understated."

When the Stork goes to the negative end of the spectrum, one hears staff members' descriptions like this. "It's not so much what he wants to do as it is the manner in which he is doing it... You can't treat people like dirt." The same individual is described to Congress and the news media as "arrogant, profane, ill tempered, abusive, demanding to a fault and a hound for perks that befit his rank," particularly when traveling on official business. Clearly, there is not a well-formed sense of personal value without the proof of external power over others.

Foul Storks become rapacious (even though, technically, a Stork is not a raptor). S/he raids other companies, womanizes/ adulterizes, and takes anything else that s/he covets — it is simply his or her "due."

Many of the descriptive symbols for foul male Storks are military or governmental in nature — the tyrant, the dictator, the grenade, the jungle fighter. The ruthless guerilla fighter probably should be saved for the Cormorant profile because of his guile and cunning.

The Queen Bee, a metaphor found with some frequency in recent business magazines, describes a woman who sounds straight out of a fairy tale or the personification of the Wicked Witch of the North. She is a master at put-downs, but cannot laugh at herself. She is particularly enraged by others who seem to be successful with a different behavior pattern and will attack what she sees as their manipulation to get what they want. Jealousy is the key motivator here. She wants no other women, specifically, in her purview, and if one follows the metaphor to its conclusion, she consumes the drones who work for her.

The equivalent in the male world may be the king of a highly specialized and perhaps secretive domain. His intellect is without question, and yet he is not supportive of potential successors.

The foul fowl is the parent who teases his or her children with a hostile humor and gives love conditionally. Only if the child obeys unconditionally will s/he be safe from certain outrages and uncertain eruptions that probably have nothing to do with his or her behavior. This is done unconsciously by the Stork. At such point as the behavior becomes conscious intent to hurt or abuse, then the Stork has become a Cormorant.

Robert Coles, in his book *The Moral Intelligence of Children,* described a child who "bragged, strutted and pushed others around and showed himself capable of deception, callousness, and brutishness. He was also a forceful, shrewd, even charismatic person, able to direct and command and persuade some, while putting others decidedly off.....with combative assertion to convey the importance of one's own sovereignty." What does the future hold for this child Stork-becoming-a-Comorant?

STRATEGIES FOR INTERACTION WITH A SENIOR STORK

Susan Brownmiller, in *Femininity,* describes the work of zoologist Thelma Rowell:

Rowell postulated that a dominant primate might be defined as "one who did not think before it acted" in encounters with others. A dominant animal walked down a path or sat on a log where it

pleased, without acknowledging a subordinate's presence. Subordinate animals cringed or scattered at the other's approach. "Thus," she wrote, "it was the subordinate animals that cautiously observed and maintained the hierarchy." Rowell suggested that "the subordinate's behavior elicits dominating behavior rather than the reverse" in many situations.

Because the Stork is so determined to win, the best strategy is to find a way for him or her to win while you also win. The win-win approach is still the superior approach, but it must not look like a compromise to the Stork. Often, the senior Stork may be belligerent over the phone or by e-mail, where s/he does not have to see you in person, especially around an issue on which s/he is not an expert. However, if you will go to meet him/her reasonably, with your facts in hand and your behavior non-confrontational, you will probably be able to achieve a win-win.

If you must brief a senior Stork, do so briefly. Use a sequential order for the information. Have back-up materials ready in case the Stork questions your analysis, but do not offer it until queried. Flow charts are appreciated, but the key word is "flow." Avoid the classic bean-counting approach; provide only the broad brush until otherwise directed. Make an appointment to see him/her and expect to be kept waiting awhile. Appear interested and alert to office movements while you wait, camouflaging any annoyance you may feel.

Another important strategy with a combatant Stork is to position yourself well at the table. An experience with a Stork over the phone on a very sensitive corporate issue indicated that one should not sit across the table from him in person. By moving to his right side, where he would have to deal with his perceived challenger with his left brain, a psychologist consultant was able to diffuse his global anger and work with him on a win-win strategy for all concerned. Interestingly enough, the same energy that had been exercised in anger became focused on helping one of his employees to find appropriate mental health support.

The combative senior Stork may be an individual who has been poorly supported or mentored in the workplace. One such Stork, after delivering a very mean-spirited Friday afternoon

pronouncement to her small senior staff with whom she had worked for only six months, was surprised that her "love me or leave me" theme produced just exactly those results. Her most able employees departed within a month. She was truly astonished and said, "I only meant to get your attention. When you are boss someday, you will have to do the same." Those who departed believed that they would be boss someday and would never be that foolish.

Often Storks do not want to work with other Storks. They will purposely seek out others in the other Stork's staff with whom they have a better interface. This puts the chosen staffer in an awkward spot. The best strategy, if you are the chosen interface, is to be upfront with your Stork. Clear any decisions or positions with your Stork in advance and report the outcome immediately. While your Stork may be offended on the surface, s/he may be delighted by the face-saving mechanism and s/he can always blame you if something goes wrong.

Storks have a tendency to use foul language. After a lifetime of such behavior, they may indeed use it unconsciously. However, it can be used consciously to intimidate others. While often extremely difficult to do, it is important to remind a Stork that you find the language or comment offensive. Unless you do, you become a co-conspirator in his or her mind.

The same strategy should be used when a Stork treats you in a way that you find objectionable. While you must be very strategic in your letting the Stork know, failure to do so will enable a pattern to develop and it will get harder and harder to step up to the plate. Let the Stork know that while s/he may be trying to help you be more productive, effective, or whatever, his or her anger or other behavior causes you to shut down. People cannot be truly productive in an environment of fear. S/he wins because you fear him/her and s/he needs the daily boost that comes from that; but, you have also won, because the behavior will probably stop or stop more quickly with one look from you the next time. Another technique for dealing with these explosive types who are senior to you is to interject with something innocuous like, "Excuse me, Sir..." This shows respect, but interrupts the volcanic eruption and

may allow you to help the Stork diffuse and refocus. If the eruptive Stork is your peer or junior, by all means, disrupt the flow.

Sometimes the behavior is not awful, but signifies something that you find offensive. For example, one senior woman told about her boss calling her "Mom." While he probably did it with a certain sense of humor, it had a demeaning ring to it. She also wondered what his interactions had been with his mother. So, she forcefully, but politely, said, "But, I am not your mother."

A foul female Stork once indicated to her female boss that her leadership felt like the Stork's mother. The same issue is at stake in both scenarios. Those who are reminded of their mothers own the problem and should not carry it over into their workplace.

A Stork demands unconditional loyalty and obedience. S/he does not understand the concept of having to earn that loyalty (or respect), however. Nor does s/he necessarily provide unconditional loyalty to those junior to him/her. For those up the chain of command, s/he will be assiduous in watching their every wish and will execute their perceived commands, even if the seniors have never indicated such a request. Junior Storks can put seniors in very dangerous or unpleasant circumstances by seeking to please without permission.

STRATEGIES FOR INTERACTION WITH A JUNIOR STORK

Junior Storks prefer discrete tasks for which they are responsible and in control. A supervisor will get the best results from a Stork if milestones and briefing points are set at the beginning of the task, and access is encouraged if the Stork runs into any unanticipated obstacles. If a Junior Stork receives an assignment that frightens him/her because s/he does not know an immediate solution, the supervisor needs to help structure a possible solution and praise him/her when it is completed. If the Stork manages a small team, the supervisor should keep an eye on team morale, just in case a mutiny is in the making. A stressed Stork can make life difficult for his or her reports and then preen when successful, as if no damage were done to others in the interim. A successful Stork can be a charismatic leader.

If you are dealing with a junior Stork who is misbehaving (e.g., one who never does his or her homework on time, or leaves his/her metaphorical dirty laundry strewn around and is acting out his/her aggressions), then it is important to hear this Stork's complaints, acknowledging his or her perspective, and together, work to come up with a satisfactory solution for all involved. You may have an independent thinker with backbone. However, if your attempt to tutor this employee does not achieve a change in the Stork's behavior, then get rid of him or her! A great deal of energy can be expended by well-meaning bosses to try to make dysfunctional Stork employees functional. But, others whom you value are watching. A good faith attempt is all that is required.

Sometimes a junior male Stork will attempt to intimidate his female boss by standing at her desk, as opposed to being seated. If he does not accept her invitation to be seated, she must indicate that he will sit or she will not listen to what he has to say. The action of sitting down provides at least a brief interlude and changes the physical exclamation point to a seated question mark. If he refuses to comply, she can either call in witnesses or leave the room herself. Clearly, this individual has become a foul fowl. Counselors know the rule to sit closest to the door of their room so that they have easy egress in the case of volatile clients. Unfortunately, in today's more volatile workplaces, the same wisdom applies.

A junior Stork becomes a foul Stork when he purposely and maliciously intimidates co-workers by flexing his brawn, getting in their face/space, or by showing a menacing, cold demeanor.....and then is foolish enough to be overheard by his supervisor bragging to his buddies about his prowess. Removal needs to be swift and with no mincing of the reason for dismissal.

Within safe settings, some Storks will give co-workers or their spouse a key phrase to use to when they are out of line. While that may seem like an attractive option, the difficulty is that the co-worker or spouse now has responsibility for the Stork's behavior. It can look like a co-dependency for an addiction. The addiction here is power.

When cornered, a female Stork may cry. When she regains her composure and has a chance to ruminate on her experience, she will be angry and will need to get even.

When a Stork realizes that s/he may have taken on too much responsibility for others' safety, s/he may be very vulnerable to relying on others who will manipulate him or her. Cormorants particularly watch for this opportunity. The Cormorant actually fears the Stork, but likes the Stork's protection and power, so they will collude as a potent duo. This leads to a foul situation. The only solutions are removal of the Cormorant, preferably with replacement by a strong, positive bird who can give the failing Stork sufficient support to be his or her best self.

BOTTOM LINE

Exposure to a foul Stork for any extended period of time is not good for one's physical, mental, or spiritual self. One may become physically sick, psychologically stressed, or compromised at the moral or spiritual levels. If the foul Stork can be removed from the environment, such as by a vote of church members, then that would be optimum (though not easy). If the foul Stork is too firmly entrenched, then you must make the decision to leave for your own well-being (also not easy, but a critical choice for the rest of your life).

Plebes at West Point must learn the following quote from General Schofield, written in 1879:

> *The discipline which makes the soldiers of a free country reliable in battle is not to be gained by harsh and tyrannical treatment. On the contrary, such treatment is far more likely to destroy than to make an Army. It is possible to impart instruction and to give commands in such a manner and such a tone of voice to inspire in the soldier no feeling but an intense desire to obey, while the opposite manner and tone of voice cannot fail to excite strong resentment and a desire to disobey.......He who feels the respect which is due to others cannot fail to inspire in them regard for himself, while he who feels, and hence manifests, disrespect toward others, especially his inferiors, cannot fail to inspire hatred against himself.*

Probable Storks that I know:

____

____

____

____

Snowy Egret

Snowy Egret

The Snowy Egret, found almost everywhere on the North American continent at one season or another, is a medium-sized bird of 20 - 27 inches. All of his feathers are white. This Egret's most outstanding feature is his head plumage which stands up and spreads out like a fountain when he is excited (usually angry about an intruder in his favorite fishing spot). He is a beautiful bird, frequently preening his much - sought - after feathers. Plumes were the fashion in the late nineteenth century, and traders captured this species almost to extinction.

He has a slender black bill topped by a yellow band of bare skin up to his yellow-ringed black eyes. Because his black legs are common to the immature stage of other species, it is important to notice his bright yellow feet. When flying, an Egret tucks his neck into a curve. When hunting, he stretches his neck up to increase his sight line and striking power.

Feeding occurs in shallow water or along the margins of fresh and salt water marshes, man-made roadway dikes, the beach at Sanibel Island, and in ponds and rice fields. His primary foods are small fish, amphibians, worms, crustaceans, insects, and sometimes small songbirds.

FOWL FOLKLORE

Pygmalion became convinced that human women were abhorrent, so used his gifts as a sculptor to create the perfect woman in ivory. She was so beautiful and his art was so great that he fell in love with his creation. Venus, knowing his desire, granted life to his statue and a happy union. Fairy tales have repeated this theme as has the popular musical, "My Fair Lady."

In the Bible, one sees the Egret Pharisees. Clearly threatened by a gifted youth who seemed to have better, or at least different, interpretations of the Old Testament stories and thorny religious dilemmas than they did, they quizzed him for many days, hoping to trap him in heresy. They knew that his ascendance would be their downfall.

In the Biblical story of Mary and Martha, Jesus responds to Martha, who has criticized Mary, that Mary's behavior is to be admired. Mary has chosen not to follow the traditional role of food preparation, and has, instead, elected to sit at the feet of Jesus and listen to his stories. Martha, the Egret, who is busy seeing to all of the details of having unexpected visitors for dinner, is much confused by Jesus's approval of her sister's behavior.

The Lakota Indians described Summer as a mouse, a worker and moderator who can get too wrapped up in detail.

There are a number of Stork-Egret and Pelican-Egret combinations in popular plays and television shows. Two that come to mind immediately are "The Odd Couple" and "Magnum, P.I." Obviously, Felix and Higgins are the Egrets; Oscar is the Stork and Magnum is a Pelican. They need each other, but antagonize each other at the same time. Marian the Librarian in "The Music Man" is wooed and won by a Brown Pelican, the spell-binding instrument salesman who followed none of the rules for teaching music.

OBSERVED FOWL BEHAVIORS

When feeding, the Snowy Egret has at least five different modes of behavior. He can feed by chasing schools of minnows and shrimp through the water, almost like a short-distance sprinter. More common is a slow walk through shallow water, wiggling a yellow foot to stir up food.

The Snowy Egret is about the same size as the White Ibis, and not only tolerates his presence, but may seek out a small group of Ibis with whom to feed. Because the Ibis are probers and are seeking different food, the Egret benefits from what they have stirred up. Only when another Egret attempts to join the group will he become feisty.

Egrets will feed in groups when the food supply is large. When the food supply is more spread out, you will see Egrets either alone along the beach, wading just in the wash of the tide, or spaced at almost exact distances of about 30 feet from other Egrets. They are not working the beach together.

On the beach, he will tilt his head at an angle to see into the incoming waves. At other times, he will wait with the patience of Job, peering into the water from a marshy bank or culvert embankment, waiting for just the right prey. He stabs with great speed, and should he miss, will utter a low croak and shake his head, as if scolding himself. With his white under side, he is hard for his prey to see against the sky.

A fourth option, seen only when the volume of fish in a dike was enormous, was a short flight with a dip from the air to snag a fish and then flight, with a foot dragging in the water, to the bank on the other side of the dike. Both Snowy Egrets and Great Egrets joined in this incredible feeding frenzy, going back and forth across the dike, re-starting their pattern after consuming their catch. There were 50 - 75 birds involved in this choreography. They were noisy, but not territorial, because there was so much food.

A ranger at the "Ding" Darling National Wildlife Refuge also reported a fifth feeding behavior. Rarely, several Egrets will work in tandem to encircle a small school of fish. When they have herded them into a small enough area, they all pounce and feed.

At other times, when there are small gatherings of Egrets at the fishing pier near the Sanibel Lighthouse, they are very territorial, chasing others away by flapping their wings and making noise. They do keep company with Cormorants at the "Ding" Darling National Wildlife Refuge, especially at dusk as they settle on small islands for the night. Not many other species keep company with Cormorants.

A Snowy Egret is not intimidated by larger birds. In an area where fish had congregated at low tide, one little Snowy Egret was in the midst of lots of relatively huge Great Egrets who squawked at each other to keep their turf. When the big ones tried to intimidate the little Snowy Egret, he fluffed out all of his feathers in quite a display. And, he was aggressive enough to continue that behavior each time he chose a new location.

Even though the Snowy Egret is very skittish and wary, he will overcome his natural hesitancy and hang out with a Big Blue Heron. They are after different food, so the Egret benefits from what the Big Blue Heron stirs up but is not interested in.

OBSERVED HUMAN EGRET BEHAVIORS

The Snowy Egret is frequently described as a perfectionist and one who knows everything. In moderation, these are admirable traits. In the extreme, they can serve as a distraction from tackling more important tasks that may frighten the Egret. The Egret's most important attribute is his/her integrity. Usually, an Egret finds a niche that allows success in controlling often-repeated processes or an intellectual domain over which s/he has encyclopaedic knowledge. These can range as widely as chemical production and engineering projects to the more common workplace administrative and accounting tasks. Inspector General personnel can fit this profile. The only caveat on his/her appropriateness for a position comes with any need for contingency action requiring immediate response. Not a risk taker who stretches his/her neck out, nor one who improvises, an Egret can be a good manager or inspector, but usually is not a leader. The human mimics the fowl by literally carrying his or her head in an upward-tilted direction.

Many bureaucratic regulatory commissions would benefit from an Egret. Any agency that has oversight of immigration, naturalization, taxation, or examination of records for government or military service needs the thorough, detail-oriented intellect of the Egret. A literal, concrete thinker, s/he is not bored by repetition; instead, s/he finds it comforting and is sharp at recognizing when there are discrepancies.

An Egret can be an absolutely dependable member of a team and often is highly valued as a junior member. When given tasks, s/he will complete them, perhaps in more detail than necessary. S/he is an introvert who likes a hierarchy. S/he does not necessarily want to be boss, but wants to know exactly what is expected, how s/he will be evaluated, and how much control s/he has. S/he is not a natural leader. S/he does not feel a real ease with people. Others will sense a distance that is purposely maintained, whether because s/he is innately shy or because of a sense of superiority, privacy, and boundaries of all sorts. Not wanting to miss others' interactions, s/he will stretch his/her neck to be able to see more. Direct eye contact is usually averted, but there are instances of bold, aggressive Egrets who practice a piercing eye contact for effect.

The Egret has worked hard to create for himself/herself a known world. Values, principles, and rules are absolutes in the Egret's world. It is hard for an Egret to take pleasure in what s/he has accomplished. Everything is polarized: good or bad, right or wrong, neat or messy. There is no in between. Order and sequential processes are requisite for the Egret's comfort and optimum performance. The philosophical and spiritual base from which an Egret operates is the perfectibility of human nature and processes, which assumes their current faulty state. S/he is a bona fide pessimist with a drought expected to diminish her already half empty glass. The Egret has layers of fail-safe plans to preclude disaster.

Incredibly self-disciplined, an Egret must keep watch all of the time to ensure that s/he does not make a mistake, nor do any others. And, it is necessary to point out those mistakes. An Egret feels compelled to criticize others, often publicly, regardless of their seniority. Whistle-blowers are usually Egrets. Childhood versions are tattle-tales. S/he is a self-appointed conscience and keeper of values (even if they are self-selected values). S/he really is not interested in the impact remarks have on others, except to motivate them to fix the problem. This fowl eats small songbirds; i.e., s/he kills the small delights that make a difference in people's lives by his or her incisive, insensitive insistence on rectitude.

Seniors find this officious and avoid the Egret whenever possible. An Egret likes to be aligned with a Stork, but is disillusioned when the Stork departs from the rules to pursue self-interest. However, because s/he lacks the power to combat the Stork, s/he snipes. So, an Egret often finds that s/he is not selected for special teams or for key positions, for which s/he thinks s/he is eminently qualified.

Like a Stork, the Egret is not malicious per se; s/he is either oblivious to the personal hurt or expects one to "get over it." If the dynamics are such that a junior could establish a mentoring relationship with the Egret, that is a positive option. The teaching dynamic might soften the martinet (or martinette). Friendship for an Egret is really a superficial connection; most folks are kept at arm's length because s/he treats his/her own inner self this way.

It is important to distinguish between the motivation for perfection by an Egret and a Cormorant. For an Egret, perfection is his/her end-all or be-all. It is the only way to do something. For a Cormorant, the search for perfection is motivated by the desire not to get caught. S/he seeks to catch others ("gotcha") in not being perfect and then cut them down to less than his/her own size. Egrets need to be wary of Cormorants, because even Egrets cannot be perfect at everything all of the time. For example, when an Egret-becoming-a-Cormorant has enjoyed a lovely dinner with old friends who have supported her through major crises, she comments to the hostess as she leaves that heated plates would have been better. And later, she relishes telling that the hostess heated her plates the next time. She was lucky to be invited the next time!

An Egret will focus on quantity, because it is measurable; often, quality is a harder construct. An Egret's focus will be almost entirely upon his/her division or department instead of an interest in how its work meshes with the broader corporate mission. Networking is not an admired skill.

An Egret likes edicts. S/he can be a PETTY TYRANT (and one needs to accentuate both words). Environments in which an Egret can thrive are large beauracracies with tomes of rigid human resource and process manuals. Everyone receives the same treatment. There is no tailoring to meet the gifts or needs of specific employees. In a Beetle Bailey cartoon, Sarge is retiring. As he looks at the geese in the sky and sees one out of formation, he points his finger and says, "Get back in line!" Egrets make good drill sergeants.

They also make efficient nurses, but nurses you might not welcome when their rules preclude giving pain medication in a timely fashion. The same Egret nurses will awaken you when you have finally fallen asleep, because it is time to perform certain protocols required by their schedules.

The Egret can access already existing information quickly. S/he takes pride in having a good memory for incredibly small details, as well as an enjoyment of computerized information systems, files,

folders, notebooks, and, the omnipresent notebook or electronic equivalent for taking notes and recording ideas. Being able to be reached at all times is very important to an Egret, second only to being able to reach one's peers or subordinates instantaneously. Not all of the cell phones and pagers that you see in meetings and airports belong to Egrets, but a fair proportion do. An Egret cannot relinquish the umbilical cord to the office for three reasons: (1) to be perceived as needed and powerful; (2) to check on whether subordinates are doing what they are supposed to be doing; and (3) to be able to get information so that s/he will not look foolish.

An Egret is very interested in the source of information. Conversations with an Egret can often feel like one is being interviewed for publication. The intent usually is not malicious, but it feels like an interrogation. Experts are people who live far away. It is very difficult for an Egret to believe that someone close at hand could be of national renown.

An Egret does well on the fill-the-blank history test, but struggles with the comparative essay. An Andy Capp cartoon captures the intellectual pattern of the Egret when the bartender says, "Actually, he's really shallow. He just makes the water muddy enough so that folks can't see the bottom."

Calendars are really important to an Egret. There is a tendency to take one's calendar or electronic version to all meetings, so that one can schedule anything that comes along. Other birds avoid this in order to determine what they really want to attend, which annoys an Egret greatly. His or her watch beeps at key times to ensure that s/he is prompt for a meeting. (Other birds use their watch alarms to curtail meetings or provide an excuse to leave.)

An Egret is conscientious and often a workaholic. The details become overwhelming and the inability to prioritize tasks leaves the Egret swamped. S/he is then aggravated if subordinates do not put in the same amount of time and effort. Egrets may well make up the majority of the sufferers of chronic fatigue syndrome (CFS). A longitudinal study, reported by Helaine Olen in Working *Woman,* found that managers "characterized as perfectionists (by themselves and their co-workers) had a 75% higher incidence of illness than their better-adjusted officemates." They believe that by

scheduling everything to a fare-thee-well, that it will all get done.

Form matters more than content. Rote matters more than reason. Decision-making is very difficult for an Egret. S/he is never sure if all of the information is in. S/he worries about being incorrect. Major General Perry Smith, in his list of "38 blazing flashes of the obvious," shared the insights gleaned from observing the most senior officers in our armed forces. Flash number seven: "When top leaders have about 60 percent of the information that they think they need to make a prudent decision, they must then decide. Leaders who demand perfect or near perfect information are usually months or years late making decisions."

Shopping is likewise difficult and generally not pleasurable. Whereas other birds will check out "Consumer Reports" or the equivalent in the situation and then proceed to look at a few options, the Egret will feel that unless s/he investigates every possibility himself/herself, then s/he has somehow cheated and deserves to have less quality than if s/he had seen six, bought three, returned two, and then wondered if the same store s/he visited earlier that day might have put out new stock by now. If you are a real estate agent, you might want to stay away from an Egret client. S/he will see every flaw in a property, where other birds will see possibilities. You will more than earn your commission. On the other hand, you may be able to sell the same house a year or two later, because of the Egret's restlessness and inability to find the "place just right."

In spite of an Egret's reluctance to close the decision door until absolutely necessary, s/he does know everything. His or her opinion about everything is correct. S/he simply knows it. Contradiction, correction, or an attempt to cajole (unless you are a favorite) is not to be tolerated. S/he says things first and thinks second. In this politically correct era, the Egret is busy stabbing prey.

Egrets will often volunteer to serve on town committees that make critical decisions regarding zoning, planning, environmental issues and historic districts. Enamored of the rules and their power to enforce them, they sometimes get carried away with their inability to tailor decisions to situations.

The Egret is persistent and believes that success will be the result of perseverance. This is the Puritan work ethic with the twist that only those who work hard will produce a noteworthy product. When others, with a different paradigm, accomplish noteworthy products, they are "lucky." While persistent, the Egret does not demonstrate energy.

VISIBLE SIGNS

The Egret usually is very self-controlling and is very thin. Some adolescent girls become anorexic or bulemic in an attempt to control their own bodies. Their families of origin often contain a female Egret. The Egret's hair is often rather wispy. S/he spends more time than other folks preening, with a lint roller being a favorite tool (often used on others, as well, who do not live up to the Egret's level of pristine presentation). While some might interpret this very personal nit-picking as motherly, that motivation belongs to the White Ibis. A high profile Egret appears as if ready to be on the cover of a fashion magazine. Everything matches and is arranged just so, but it is black, Navy blue, grey, or another neutral color. Other Egrets are more no-nonsense in their apparel, but always neat. The character of Phyllis Lindstrom on "The Mary Tyler Moore Show," is a classic Snowy Egret.

The Egret's office will be neutral in color and showcase only the requisite credentials and awards. His/her office drawers or shelves will be spartan, with only the references and manuals necessary to accomplish his or her tasks. There is a plain mirror to check one's appearance before departing the office. There are few clues to hobbies, family, or interests beyond this job. Any art will tend to have a moral story, sometimes with a moral text incorporated into the design. Projects will be stowed in discrete folders or piles. Woe be it to an assistant who straightens the piles! There will be distinct protocols for incoming and outgoing materials. Accountability and locatability are key.

The Egret has very good handwriting. Sometimes it is the classic cursive writing of teachers and sometimes it is very precise, small handwriting. It usually accomplished with a favorite pen,

often ink as opposed to ball point, with a very narrow point. An Egret's signature is clearly legible.

When an Egret has a hobby, it tends to be something that requires great attention to detail (such as English bell ringing or historic model building) or self-discipline (such as a demanding physical fitness regimen). An Egret can be so focused on being thin that his or her fitness regimen actually causes damage to his or her body.

HUMOR

An Egret's humor can be waspish, stinging, acerbic, and petty. It is designed to diminish others. There is also a delight in toilet-bowl humor. In this era of politically correct conversation, the Egret can be a real embarrassment to cohorts, getting off on a tangent of remarks with blatantly sexual overtones. The Egret, who usually is publicly prudish, sees himself/herself as immune to the requisite niceties. The Egret is particularly nasty in regards to others of different ethnic, religious, or class affiliations.

POLITICS

In politics, an Egret will memorize all of his or her speeches and any supposedly ad hoc comments, giving exactly the same stump speech several times a day, with the same pauses and inflections. Nothing is spontaneous, because s/he has learned that s/he gets in trouble with shoot-from-the-hip decisions or comments. An Egret prides himself or herself on being a "straight shooter," honest, without guile, and no pretense. "Truth is the best defense," but truth is determined by the Egret. S/he is duty bound, and will mention duty often in speeches, including the fact that s/he calls an elderly parent or spouse daily while on the road. S/he must be in control of all interviews and situations, so questions from the press must be provided in advance. Photographs of some current Egrets show them shaking their finger at their opponents in national debates, at members of the press corps who have challenged them, and in campaign speeches. The preacher is a scold. Unlike the Stork politician, who likes to give others nicknames, the Egret will not tolerate a nickname.

FAIR FOWL

A fair Egret may be an incredible combination of beauty and brains. Immaculately tailored in dress and behavior, s/he does not seek the limelight. In fact, s/he is uncomfortable with marketing his or her work and must expend significant emotional energy to have a visible profile (often necessary for entrepreneurs in fields like financial management and teaching).

A fair Egret is precise in planning events and is a superb logistician. S/he establishes protocols, check-off lists, and after-action reports that enable others without his or her organizational talents to perform complex, detailed tasks. S/he is especially effective in a team with a creative co-chairman. When both appreciate what the other brings to the mission, this is a dynamic duo.

Egret accountants can seem extremely divorced from the organizations they are called upon to serve. When fair Egrets are involved, they become involved to a degree sufficient to allow them to counsel those they serve. Professional ethics and boundaries must be observed, but as any leader knows, "figures lie and liars figure." It really helps when fair Egrets can become more than bean counters and begin to assist in the overall complexities of judgment that leaders must make.

The Egret can demonstrate a delightful sense of humor away from his/her workplace and with people with whom s/he feels comfortable. Fair Egrets sense that they need to surround themselves with people who bring out their wit and delight, and that they need to balance their formal work with hobbies that allow them to express themselves artistically. One such Fair Egret gardens extensively, but with banks of wild flowers instead of the more characteristic defined beds.

A fair Egret piano teacher, who would love to have her students perform well at their recital, has the grace to say to a student, "Debussy would have liked that ending, too."

A fair Egret manages his or her needs for perfection by working solo for a major portion of the day. Having gained sufficient control and calm in this environment, s/he has sufficient reserve to

participate in the melee that characterizes many workplace meetings. When given an assignment that may at first seem beyond his or her grasp, s/he will tend to believe that this will be his or her downfall. However, if a supervisor is supportive and addresses some of the small steps necessary to get the project started, the fair Egret will regain self-composure and begin to feel successful as the small first steps are completed.

When a fair Egret queries you with his or her very thorough questions, you usually will feel pleased by the attention and concern.

When the Internal Revenue Service audits a fair Egret's returns, they find meticulous notes and receipts, and often over-payment. The fair Egret bends over backwards not to break any regulation and in so doing, makes the regulations more definitive or damaging than originally intended.

Courageous Egrets often become whistleblowers. While usually working in the lower or middle echelons of a company, they hold positions such as accountants or lawyers — positions from which they may notice irregularities (as in the recent WorldCom and FBI revelations). They are dogged in their determination to spotlight institutional flaws.

A fair Egret may willingly take on family genealogy tasks, loving the history and the detail. An Egret can be a fine scientific researcher, enjoying the detail and privacy.

Service-oriented professions benefit from fair Egrets, who have not only the exquisite attention to detail required in airline service, dental hygiene, and banking, but also have the interpersonal skills to focus on the customer.

Sometimes, superiors can mistake a fair Egret for a Blue Heron because a healthy Heron has many of the perfectionist qualities of an Egret (but is focused on high priority tasks only). A fair Egret differs from the Blue Heron in another way; s/he tends to do most of the work himself or herself (and can become overwhelmed in the process). Whereas, a Blue Heron is more collegial and collaborative.

FOUL FOWL

The Egret is at risk for burnout or at least a martyr complex. Driven by an overwhelming compulsion for perfection in all things which, of course, is mortally impossible, and having alienated folks who might have been inclined to be supportive, the Egret really is alone in his or her misery. An Egret is often in the workplace with Storks whose energy and demands can be overwhelming, and they do not notice their impact on the Egret.

Unhealthy Egrets suffer from stress-related conditions such as ulcers, arthritis, and brittle spine. Their voices are often chronically hoarse, from overuse and tightness. They scowl.

The foul Egret has secrets, but they are Secrets. Because s/he strives for perfection, a perceived flaw must be covered up. This can be one "big" flaw, such as age, education, attainment, job experience, etc.; but in order to camouflage it, a web of lies must be constructed and protected at all costs. It becomes the Scarlet Letter. Woe unto anyone who discovers or divulges — woe and wrath! The irony is that the flaw is often a molehill in others' eyes.

The foul Egret has become the world's policeman. S/he rules her fiefdom like a martinet. S/he is not concerned with his or her impact on individual people; s/he works for a principle or abstract value. The impact of the Egret's endless rules on others, according to Philip Yancey in *What's So Amazing About Grace?,* is that they learn to rebel......to feel a constant temptation to resist the demands of the authority simply because they were imposed as demands."

The workaholic boss or entrepreneur can create a work environment that is toxic to the health of his or her employees. Expecting others to mimic his or her style or drive may lead to their feeling that nothing they do will ever be good enough. Even healthy Egrets do not praise their peers and juniors enough. Imagine the dynamics for the unhealthy Egret who has a tendency to believe everyone is out to sabotage his or her success. They may be.

When a foul Egret encounters an extremely able employee (or child), then his or her strategy is to demand perfection, thereby whittling the upstart down to size (i.e., less than s/he the Egret is). Fortunately, foul Egrets do not survive long in competitive environments. Superiors notice that they are causing hate and discontent and that their micro-vision is stalling completion of tasks. Superiors also notice that star performers leave an Egret's domain. Either the Egret moves on of his or her own accord, or s/he is shuffled to another division or solo position where s/he cannot cause such havoc.

STRATEGIES FOR INTERACTION WITH A SENIOR EGRET

An Egret's propensity to inspect is a sign that s/he expects that others will not perform without his or her oversight. S/he insists on uniform procedures and distrusts flexibility, innovation, and risk. S/he sees uniformity as an issue of integrity; others see it as rigidity.

Because a senior Egret values perfection, it is critical that any data submitted be absolutely correct and that the form of written materials be grammatically correct. This is requisite before s/he can begin to assess the quality of the content. Rough drafts need to be smooth, because the Egret is quickly distracted by little annoying errors. A senior Egret administrator inspects every detail, but rarely prioritizes tasks. Everything is important to an Egret. S/he sees the trees, but not the forest.

If new data becomes available, it is helpful to insert it in already-presented materials in a format that highlights what is new. A jumble of new information poses problems for an Egret, who does not even like to make a rough draft with fill-in-the-blank spaces for answers as they become available.

The Egret must win, so the most a junior can hope for in a discussion is a win-win. An Egret may ask you the same, truly identical question, fifteen times in a row. While some might interpret that as trying to grasp the complexities of an issue, which often is difficult for the Egret, some feel the inquisition is more directed at determining if you will break down and give a different

answer. This is the patient Egret that finally stabs the fish.

In order to avoid being eaten for dinner, the junior must recognize the senior Egret's compulsion for flawless performance. If the junior makes a mistake, s/he should admit it promptly, indicating lessons learned and provisions now in place to prevent a repeat.

In order to cope with the compulsion to have assessed every option for a given decision, a junior may suggest to a senior Egret that a company or person s/he admires selected a given product. Especially when the decision is complex, the potential for an Egret to get mired in the possibilities and miss critical time lines is great. A classic family example is the wedding of the first daughter. If the Egret has admired someone else's version, then getting permission to follow the advice of the predecessor (with a few personal touches), is brilliant.

It can be comforting to a junior to work for a senior Egret, if s/he goes to bat for the junior who is being badgered or treated badly by others. Aware of all of the rules, the Egret will take on anyone whose actions are out of bounds, not out of a sense of compassion, but out of a sense of regulations.

When interfacing with an Egret, it may suit a junior to flatten his or her profile as much as possible and focus the attention on the Egret. An Egret really is not interested in hearing about your dilemmas. Only his or hers count. By ensuring a high quotient of self-esteem food for the Egret, you may help your Egret perform optimally.

Another reason to flatten your profile is that an Egret, like a Cormorant, feels a discomfort or irritation if your individual work is noted by higher ups. It feels like one-ups-manship to the Egret, even if you did not seek the visibility. The result will be tieing you to your station (literally and figuratively). You will find yourself left out of key meetings and may be exposed to his or her sniping in front of your co-workers and those more senior. This comes from the Egret's sense of there being a finite supply of resources, praise, and power of all sorts. You may be getting some of the Egret's share.

For the same reason, avoid taking public credit for the work you have done (though you should document it and ask that it be included in your annual evaluation). A female Egret sees this as bragging and especially inappropriate for a woman. This individual prefers no visibility for herself as well, because she fears failure with the very next wave.

Keep a list of the tasks assigned to you by your senior Egret. Request prioritization. Carry the list with you when you meet so that you both can go over its current status. Some senior Egrets will make that list for you on shared computer links. While it may feel punitive, it is a way of keeping track for both of you.

In order to give your senior Egret a sense of comfort on the progress of key projects, arrange to update frequently. Check with him or her to see if s/he prefers that you simply drop by or if s/he would prefer prior notice. Some Egrets feel uncomfortable with an open door policy and dislike being interrupted when they are focused on a task. Some will not even look up or acknowledge your presence until they have finished.

A senior Egret may flaunt the rules that s/he quotes to others, with the unspoken stance that rules are made for peons. This is the senior supervisor who requires that everyone else's time card reflect actual time on duty and that the in-out board reflect their location (if, indeed, permission to depart the building is not demanded). However, s/he does neither of the above in relation to his or her supervisor and is nasty when called on it.

In the domain of omniscience, when the Egret pronounces the proper assessment of all situations, a junior has to find a way to assert a different point of view when it really matters. When the attacks become personal, then the junior should have some strategies up his or her sleeve. For example, with a particularly difficult Egret, who is probably on the verge of becoming a Cormorant, one may set the limits that will be tolerated. One daughter with an elderly Egret mother has established her date of arrival and her intended departure date, but is clear about leaving sooner if her mother behaves badly. And she is ready to do so on a half hour's notice.

In a scenario with a former Egret supervisor, to whom s/he is no longer positionally obligated, a junior may elect to be very firm about no further communication, no further contact. As Egrets lose their status power and must earn respect, friendship, and support, they may need to learn some new skills.

Even in social situations, an Egret's comment is often implied criticism. An effective response is to compliment the Egret. That diffuses the intended or perceived criticism and makes him or her wonder if you understood the original remark. Then you can move on to a conversation with others.

It is difficult to be in the thrall of a foul Egret. If you cannot leave the relationship, which would probably be optimum for your mental health, a second option is to create emotional distance (and spatial distance, if possible). Intellectual awareness of how the Egret operates empowers you to smile to yourself as you note specific behaviors. Spiritually, it helps to remember that all folks are frail humans, including you, and that all of us irritate some others. Jesus prayed, "Father, forgive them for they know not what they do." If you are clear that the Egret is not consciously seeking to hurt you personally, you can forgive the Egret in your own mind; discussion of this personal decision with an Egret is not advised!

Forgiveness does not mean that you make a clean new start. Once you are vulnerable or hurt from an Egret's actions, self-preservation calls for being wary. The key is to try to be respectful and be your best self, not manipulated by responding to the Egret's behaviors. Do not expect change from the Egret.

If you have an Egret who is consciously seeking to do damage, then you have a Cormorant. Kindness to a foul Egret/Cormorant may thoroughly irritate him or her, because s/he intends to do damage. Extreme cases like this probably include a diagnosable mental illness.

STRATEGIES FOR INTERACTION WITH A JUNIOR EGRET

This is a worker that you can trust absolutely to work solo at the break-of-dawn set-up required to get an expensive coffee and baked goods bistro into operation on time each morning. S/he will remember what each customer likes to drink, and the cash receipts will always tally. It is important to praise his or her dependability and attention to the small details that make a complicated operation run smoothly. You need to be careful of making too many changes at once, because s/he thrives on continuity. You will also need to insure that s/he does not overbook his or her schedule. This Egret can become a workaholic, believing that no one else can do the same job s/he does. This may be true, but if you want his or her skills long-term, you will have to prevent burnout in this highly committed employee.

A junior Egret craves praise for having done a project "just as I had hoped," or "even better than I had hoped." It is easy to take precision and excellence for granted.

A junior Egret may display a restlessness in his or her job. S/he may feel that this position is not exactly right and not enough is done for him or her. While Egrets are not known for giving praise and support to others, they are greedy consumers of the same. A supervisor should look for appropriate little ways to note the quality of the junior Egret's work with some frequency (mindful, of course, of the needs of others in the workplace).

Robert Coles, in his book *The Moral Intelligence of Children,* described a boy "dryly, smugly preoccupied with the errors of others... to the point of vanity." His "child as scold" possessed a "jealous literalism." There is the intimation that a child (or adult) who doth protest too much about others' errors is perhaps struggling with the very same temptations. If he can stamp them out in others, he will have absolved himself. Teachers and other adults who interact with this child need to help him get the professional help he needs so that his life will be qualitatively different.

If you supervise an Egret who muddies the water daily and causes disruption in your workplace, it is advisable to let him or her go as soon as possible. In some human resource regulations, there is a one year grace period during which you may release an employee for lack of interpersonal "fit" in the office. The relief to the rest of your staff will be palpable. An enraged Egret can be very aggressive. S/he can put your whole office at risk.

BOTTOM LINE

Judy Galbraith, in her book *You Know Your Child is Gifted When...,* addresses the myth that perfectionism can sometimes be a good thing. Her response:

Perfectionism is never a good thing. What's good is the pursuit of excellence, which is not the same. Gifted kids (and their parents and teachers) often get the two confused. Perfectionism means that you can never fail, you always need approval, and if you come in second, you're a loser. The pursuit of excellence means taking risks, trying new things, growing, changing....and sometimes failing.

> *Once we give up searching for approval [or giving it unsolicited], we often find it easier to earn respect.*
>
> Gloria Steinem

> *If you judge people, you don't have time to love them.*
>
> Mother Teresa

Probable Snowy Egrets that I know:

Seagull

Seagull

The Seagull is one of the most common birds on the North American continent. He has adapted to a wide range of environments and has adopted some habits not common to the species in the wild. It is important to think of the wild bird in this context. The Herring Gull is the most prevalent species. He is a handsome robust bird, at 23 - 26 inches in height, with the male larger than the female (not true for many of the other Sanibel birds). His body is predominantly white with a pale gray back and feathers. His tail feathers are black and white striped. His eyes and bill are yellow, with a red spot at the tip of the bottom of his slightly hooked bill. When young nestlings peck at the red dot, the parent regurgitates food for them to eat. His legs are a light flesh color and his webbed feet indicate that he is a swimmer (mostly on the surface).

In flight, the Seagull shows his long pointed wings in his powerful flight. Some of the species soar frequently, as do Storks.

Seagulls rarely dive from the air into the water to catch a fish. Instead, they alight on the water in order to catch the fish. At low tide, Seagulls patrol the mud flats looking for food. Their webbed feet serve them well in this environment. Seagulls consume plant life, marine and animal life, refuse and carrion.

On the beach, various species of gulls cluster interchangeably. All facing into the wind and not fazed by people walking by (if a steady pace is maintained), they look like a formally dressed audience in an opera house — all in shades of black and white.

FOWL FOLKLORE

The Seagull resembles the hero's sidekick found in many companion combinations in literature. Don Quixote and Sancho Panza were seventeenth century examples of a Great Blue Heron and a Seagull, who set off on a most unusual quest. Sancho Panza, the loyal compatriot who was devoted to his visionary knight, tried to protect him from his delusions and his very real dangers. In so doing, he put his own life at risk. He delighted in his master's

dependence on him, and rarely looked after his own comfort and safety.

Natty Bumppo, the great wilderness scout, and his various Indian guides in James Fenimore Cooper's *Leatherstocking Tales* and other novels, exemplified the fearless leader and his intrepid partner. Other dyads in which the fearless hero depends upon the allegiance of his wise scout include the Lone Ranger and Tonto, Lewis and Clark and their guide Sacagawea, Perry Mason and Della Street, and Miss Daisy and her driver, Hoke, in "Driving Miss Daisy." The wilderness and corporate headquarters share this symbiotic pairing.

OBSERVED FOWL BEHAVIORS

The Seagull eats everything; he is a truly omnivorous scavenger, but generally is not a predator (though he may eat small chicks of other species). He will steal food from other birds. For example, a Herring Gull will plague a Cormorant who has surfaced with a fish until the Cormorant finally gives up his catch. Every time the Cormorant dives to escape, the Seagull watches for him to resurface to grab a short breath and try to re-position the fish so that he can swallow it. The Seagull's persistence finally wins the day, with the larger bird outsmarted and outmaneuvered.

A Seagull was observed at the "Ding" Darling National Wildlife Refuge attacking a Snowy Egret for the food he had caught and persisted until he got it. This interchange can be seen in the human equivalents.

A different relationship seems to exist between the Laughing Gull, a slightly smaller close relative, and the Brown Pelican. The Laughing Gull lands on the head of a Pelican who has just filled his pouch with fish, and snatches a fish from the Pelican's bill. It seems to be a symbiotic relationship, as opposed to the attack mounted on the Cormorant. David Allen Sibley, in his book *The Sibley Guide to Bird Life & Behavior,* calls this "kleptoparisitism."

Many of us have seen Seagulls pick up oyster shells and methodically fly off to drop them. Piers, roof tops, and parking lots are littered with the shells of Seagull dinners. In the wild, the

Seagulls are good beach scavengers, ridding the area of dead fish and other sea life, such as scallops, sea urchins, and other morsels that have been washed up and left high and dry by the tide. Where there is a good balance in nature, such as one finds at Sanibel Island, there is rarely any evidence of rotting fish or edible mollusks.

Yet another mode of food gathering was observed as a big Seagull stood at the wave line of the beach. As the wave retreated, he did a little dance with his webbed feet, opening small food to view. So, he fed much like an Ibis after stirring up the sand.

Seagulls operate solo and in groups. In some settings, where Herring Gulls have over-populated due to the proximity of fishing co-ops or town dumps, they can become predators of the young chicks of Plovers, Laughing Gulls, Terns, and Black Ducks. Then, U. S. Fish and Wildlife and the U. S. Department of Agriculture will step in for habitat control.

The Seagull is a model for diversification in the workplace. S/he enjoys mixed groups of Royal Terns, Skimmers, and other Terns and Gulls. While all are about the same size, they do not require clusters that are species-specific. Cocky, courageous, assertive, and used to getting his way, he is very conscious of the pecking order in his group (unlike the Ibis).

Seagulls communicate incessantly with each other in high squeals. Their shrill voices display a fairly complex vocabulary, albeit not particularly melodious. Their cries often sound like children wailing or screaming. They also communicate by bobbing their heads.

OBSERVED HUMAN BEHAVIORS

This is the under-sung and under-appreciated peacemaker. Others who posture as peacemakers have so many personal agendas that it is hard for them to compromise. The Seagull is truly able to scavenge among the existing morsels and find a way to agreement, taking a loss as necessary. Even though s/he resists visibility, his or her advice should be sought and followed in high profile negotiations.

The Seagull usually migrates to a position that allows him or her to be a primary assistant, executive gatekeeper, or chief of staff to senior Storks, Green Herons, or Great Blue Herons. S/he wants to be perceived as powerful, but prefers to be successful vicariously via his or her boss or spouse. S/he is a true "two-fer," a term well known in corporate and military circles. S/he dislikes a leadership role and will avoid supervisory duties of a small staff as much as possible, far preferring a one-on-one relationship with his or her boss and de facto status as the senior administrator. S/he enjoys being with a number of the other bird types, but demonstrates a special allegiance and fierce protection for his or her direct report. This dyad has a remarkable bond.

When a Seagull is the spouse of a senior corporate or military officer, s/he relishes the opportunity. Modern versions are less inclined to wear their sponsor's rank than their predecessors, and are more likely to use the opportunity to sponsor, but not lead, causes of particular importance to them. These causes are allied to and supportive of their spouses' initiatives. As such, they become powerful forces for action in domains that would ordinarily not be of key interest to their spouses, but do, indeed, positively impact public opinion. Presidents' wives have been known to operate in this fashion. They use their soaring capacity to scout for "profitable" opportunities.

When one first encounters a Seagull, s/he may be very protective of his or her lawyer or doctor boss. However, if the client interacts appropriately with the Seagull, acknowledging his or her own special expertise, future conversations will be warm and personal. When given his or her "due," s/he is a joy to be around.

The Seagull is very organized and enjoys tasks that require scavenging through sources of information to find answers. When tasked with complex projects, the Seagull works best if s/he can show the work in progress and determine that s/he is on the right track. Able Seagulls can assemble materials every bit as remarkably as Snowy Egrets. Therefore, they can be allies or enemies, depending upon the workplace.

Because a Seagull is so focused on making use of what comes his or her way, s/he rarely shows initiative to accomplish something not tasked by a superior. Disliking the limelight, the Seagull will do everything possible not to be visible. And yet, his or her commitment to the mission is without question. S/he will coach or assist others to complete their tasks, but prefers to be solo on his or her own. S/he needs to know the priorities of the boss. When a variety of tasks have been handed down, s/he does not want to make the decisions about what comes first. However, when the boss is disabled for any reason, the Seagull has observed long and well, and is very adept at covering for him or her.

A Seagull can handle chaos on occasion with good emergency response. However, s/he craves order ordinarily. S/he is proud of his or her capacity to tie everything up before going home from work... at the regularly stipulated time. The Seagull is not a workaholic.

A Seagull is an astute observer of others. S/he knows what the office politics are, even when the boss is oblivious to the undercurrents. S/he serves a critical role in many workplaces, helping the senior supervisor be aware of discontents so that s/he can head them off before they become conflagrations. Another role played is that of go-between. Many confide in him or her because s/he is truly interested in them. They forget, or knowingly hope, that s/he will share critical information with her superior.

In spite of his or her linch pin position in the workplace, people like her for herself, unless s/he feels embattled and has taken a defensive stand. Because s/he is usually an introvert and not gregarious, s/he is surprised that so many people are indeed supporters and admirers. S/he underestimates his or her ability to be valued for who s/he is.

S/he is an astute observer of his or her boss in both professional and personal arenas. A common comment is, "I read your mind, didn't I?" S/he prides himself or herself on an intuitive knowledge of his or her senior's needs, having research done, papers ready for signature, and arrangements made for upcoming events. His or her senior may come to take such excellence for granted and fail to

praise and show delight in the Seagull's remarkable support. Storks and Green Herons are more likely to be focused on themselves or on the mission. Great Blue Herons will tend to be more aware of his or her value and be grateful.

Much like the Seagull who attacks the Cormorant repeatedly for his fish, the Seagull in the workplace will attack a Cormorant. S/he often is in a position to see the malicious maneuvering going on (a la the Cormorant). If s/he perceives that it could be a threat to his or her boss, s/he is offensive and defensive immediately. S/he will also attack a Snowy Egret for food in his or her bill. The Egret is not swimming, as the Cormorant, but is equally astonished. An Egret in the workplace can feel challenged by a Seagull because of some common skills; but the Egret usually lacks the favored access to the boss. An Egret will go so far as to recommend to a senior not to promote the Seagull when a better position becomes available for the Seagull. The Seagull sees life as a "win-lose" proposition; s/he wants to win for his or her boss, but is willing to lose for herself. S/he perceives that s/he is usually on the short end of things; therefore, s/he snatches things from others.

S/he finds change challenging. When a Seagull is thrust into a situation in which s/he must depend on others, for example, in a hospital, s/he may behave badly to those who wait upon him or her. It is an expression of his or her sense of powerlessness in the absence of her boss. And, of course, it can become a self-defeating behavior, as others avoid and do as little as possible.

On the opposite end of the scale, the Seagull enjoys the Pelican in the workplace. The Pelican shares without concern, as s/he is so full of the abundance of life. The Pelican is fun to be around and the Seagull has figured out that s/he delights in the company of this ebullient fowl.

A Seagull tends to underestimate his or her ability. Not a high energy person, s/he relies on the boss to energize the workplace and s/he then responds in kind. By preference, s/he is more laconic, not given to ebullience, but able to follow the flow.

A Seagull is generally not a risk-taker. S/he prefers hobbies that allow reading, needlework, and quilting. A strong interest in history is typical of a Seagull. Biography, old movies and plays, classic operas and ballets all have an appeal for a Seagull. S/he learns lessons from these sources and they are as real as daily life.

S/he distances himself or herself from close friends in the office and has few really close friends outside of the office. S/he tends not to attach himself or herself to causes or community groups. Work is the focus of his or her life, even though there is not a significant personal achievement motive attached.

While the Seagull does not seek personal visibility, s/he absolutely thrives on praise from his or her important other. Sometimes the need for her daily portion will be couched in advice to the senior that someone else could benefit from some personal support and praise.

One sees rare flashes of competitiveness when resources or jobs are scarce. Generally, a Seagull operates from a philosophical frame of a sufficient supply of resources. Because s/he is not looking to be number one, s/he is able to focus on his or her boss and ride on his coat tails. One sees this in military assignments where the chief of staff may know that s/he has not been selected for a star (i.e., a general or admiral). But, s/he can be extremely influential and effective as the general's chief administrator. In corporate or large law firm settings, one sees the promoted senior take his trusted administrator with him or her.

Because a Seagull may be so focused on reacting to senior others, s/he may not do things that would enhance his or her own life. For example, when a Great Blue Heron would like the Seagull to "be all s/he can be," thus serving more in a partnership than a subordinate role, there is a reluctance to get the further education that would allow the superior to promote him or her. Other Seagulls show reluctance to learn skills, be they computer, finances, or other specialized training that would allow them to progress up the hierarchy or provide a safety net in times of downsizing. A Seagull can rely unduly on a senior's ability to carry him or her along.

The loyalty of the Seagull means that s/he absorbs a great deal of stress, either on behalf of or because of his or her boss. A number of studies have shown that people in the classic position of the bureaucratic assistant have a higher risk of heart disease than those whose status gives more personal control.

One of the dictionary interpretations of "gull" is one who is gullible, easily deluded, tricked, duped. The verb to gull is to delude or impose upon. A Seagull's propensity to be symbiotic makes him or her a devoted employee, friend, and spouse. However, his or her vulnerability in the physical or emotional absence of a powerful other is of great concern to those who care about the Seagull. The Seagull will tend to rely on others to advise him or her about personal finances, investments, tax deductions, and retirement plans. Even if others do not know that s/he wants that kind of personal advice and are reluctant to give the unsolicited advice s/he expects, s/he will not ask until a problem occurs. S/he trades dependency for loyalty, but not in a blatant way. A female Seagull needs to have a strong male in her sphere of influence. Conversely, a male Seagull usually has a strong female in his sphere, often an older woman or a spouse.

Decision-making can be stressful for a Seagull. If the decision or action is required for his or her boss, then s/he will make the herculean effort. However, if it is a personal decision, the Seagull has the ability to remain in limbo for what would be an excruciatingly long time for other birds. Again, because the Seagull operates from a world view of sufficiency instead of scarcity, s/he does not have the sense that failure to act on an opportunity now will mean that something similar will not come along again. A Seagull bride will change her mind about the wedding dress, the gloves, the veil and the location of the wedding, driving an Egret mother wild. But, none of the changes are malicious; she is just continuing to exercise options.

VISIBLE SIGNS

The Seagull maintains a meticulous office, often with an eclectic collection of furniture or accessories. Treasures are found in antique shops, yard sales, and church white elephant sales. His

or her ingenuity in the use of found items gives great pleasure. However, the Seagull will not hoard or stock up on office supplies as some others do. S/he is practical in determining what s/he will need and will order just those items.

House-hunting with a Seagull can be a long experience, as s/he weighs all of the options, but once the decision has been made, there are no regrets. When s/he sells his or her house, it will go quickly, because s/he has been very organized in making it ready for sale. S/he will also move efficiently and leave it spotless.

Clothes shopping with a Seagull will include non-traditional sources like thrift stores and consignment shops. S/he likes getting a bargain. His or her choices in apparel typically mimic those of the senior for whom s/he works, but s/he is practical about wearing more casual clothing on days that call for inventory or office rearrangement. Favorite colors are beige, champagne, and putty.

Some Seagulls are real puzzlers and enigmas. You find yourself going back over and over trying to put your arms around this behavior pattern. There is no malice here. But, there is a propensity to hide, and whenever possible, to avoid taking a position that can be documented. In a number two position, a Seagull may survive politically in the workplace; but if promoted, s/he may have a really hard time succeeding in the limelight.

HUMOR

The Seagull can be a very witty person. Puns, double entendres, and comparisons to famous movies or authors can be grist for the Seagull's wit. It is no coincidence that the Seagull enjoys the Pelican, whose wit can get the best of them both. Generally, a Seagull will share his or her humor only with those with whom s/he is very comfortable, not wanting to spar as some others do, and especially not wanting to be the target of others' remarks. Those who are not included in the Seagull's circle may feel a little envious and closed out. This circumspection is simply the Seagull's sense of propriety and self-protection.

POLITICS

The Independent Party should rush right out and sign up all Seagulls... except that Seagulls resist affiliation, even though they do see it a duty to vote. A more apt description of a Seagull's politics is apathy. Unless issues directly affect him or her, s/he is not likely to examine the intricacies of local or national politics. In fact, a Seagull may not even subscribe to a local newspaper or watch local cable access programs, preferring instead to read the Sunday edition of a national paper and/or subscribe to a major weekly news magazine. That way, the news can be consumed as if it were history or biography... something already completed, not in flux.

Those few Seagulls who are politically affiliated will tend to "buy the party line," questioning few of the positions taken by admired leaders. They enjoy the camaraderie of campaign headquarters, but do not seek personal visibility beyond the gratitude of the candidate.

FAIR FOWL

A fair Seagull has allied himself or herself with a few people who value him or her and encourage activities that expand life beyond the workplace and the position description. Believing that s/he is enjoyed for himself or herself, s/he is able to expand her friendships and support system. Often, others would like to be a part of his or her inner circle, but the Seagull's focus and allegiance to the primary dyad have closed off that opportunity.

If the Seagull is part of a healthy dyad, s/he may be more expansive with others because of the physical, social, or emotional support s/he receives. The aura of the powerful member of the duo casts a rosy glow by which s/he is empowered.

It is important for a Seagull to extend his or her connections so that loss of his or her job or retirement is not devastating. Even though s/he may have had many acquaintances in the workplace, s/he has had few long-term, deep friendships. A fair Seagull can be a delightful companion or friend, who puts emotional energy into pleasing important others.

An important differentiation will be identifying the bona fide Seagull (the consort spouse who focuses all of his or her attention on her husband or boss to the detriment of their children and staff) vs. the Ibis (whose role is to be the glue for the whole family or office). Lady Bird Johnson and Anne Morrow Lindbergh personify the Seagull relationship. Their children missed the attention they wanted because these high profile spouses accompanied and supported their Storks in ways that no one else could. In the wild, Seagulls form a permanent lifetime pair bond, unlike many birds who are monogamous for a season.

PSEUDO-SEAGULLS

Many social workers are trained to operate in a Seagull mode with clients, allowing clients to take ownership of their issues and, eventually, their solutions. This dynamic works well with many of the other species who want to be in control. It can be frustrating to those who want quick answers and want to be told what to do to resolve their pain, quandary, or predicament.

For those social workers and others who function similarly, who are not really Seagulls, it can be stressful to restrain their natural inclinations for so much of the day. Therefore, in non-client relationships, such as intra-staff work, they may work off their pent-up energies in forceful versions of their real selves: Storks will be more commandeering; Egrets more critical of others; Ibises more nurturing; etc.

FOUL FOWL

A dysfunctional Seagull will not be able to counter the strong influence of a boss or a mate. A Stork or a Cormorant can be particularly destructive of a Seagull in the workplace or in a marriage. It is very hard for a depressed Seagull to stand up for himself or herself in the workplace, even when all of the bureaucratic safeguards are in place. Sometimes, a Seagull will find herself in a depressed state because s/he cannot alert important others in her life to the serious hurt s/he is experiencing. S/he will be attacked by a Cormorant and not know how to respond — at first — but watch out! Because s/he cannot cry out for help when

s/he should, s/he suffers longer than necessary. Her powerful rescuer needs to be clear about the fault and blame belonging to someone else.

A dysfunctional Seagull will fly over and "shit on everyone" — a rather common, but graphic description of the Seagull's style. The angered and injured Seagull will ensure that everyone knows his or her pain.

The dysfunctional Seagull has become too dependent upon another person versus a symbiotic relationship in which there is more equality of benefit.

Sometimes a senior Stork or Cormorant will fire a Seagull loyal to his or her predecessor and then turn around and hire another Seagull, but one whose loyalty is to him or her. Sometimes that is a case of throwing out the corporate memory with the bath water.

STRATEGIES FOR INTERACTION WITH A SENIOR SEAGULL

Because Seagulls often occupy the gatekeeper position to their senior, they become "senior-by-shared-glory." The best ways to work with Seagulls is to respect their control of the keys, to formally request access with a stated mission agenda, and to be grateful for their recognition of your priority. It is critical to be honest about the immediacy of your need to see the Seagull's boss. If you do not have an emergency that the boss will also perceive as an emergency, be flexible in your appointment timing. Also, honor the length of the appointment that you are given, so that s/he can maintain the flow on schedule. Reschedule another appointment to continue, if necessary, rather than exceed. S/he has a long memory and does not take kindly to bullying, unethical runs around his or her access protocols, or failure to respond appropriately to requests that s/he has made in the name of his or her boss. In this last case, you are messing with the Seagull's ability to tie up his or her tasks each day.

The senior Seagull does cherish accurate compliments on execution of his or her duties. If overdone, s/he senses undue flattery, which will get you nowhere. Sincere appreciation,

however, is a commodity often in short supply and greatly welcome.

An incredible source of corporate history and current information, the Seagull is absolutely the operational hub of the organization. When the Seagull is absent, his or her impact is more immediate and far-reaching than any other member in the workplace.

STRATEGIES FOR INTERACTION WITH A JUNIOR SEAGULL

The junior Seagull will benefit from your accurate praise, just as his or her senior counterpart. Additionally, you should prioritize tasks that you have given and be clear about when s/he can simply execute them with no further approval necessary. This gives him or her responsibility and control which will help the Seagull become a more independent worker under sufficient direction and support.

A Seagull can grow to take on significant tasks, such as researching and compiling a human resource manual for your business. If you underestimate the ability of your Seagull to take on complex projects, then you have wasted a valuable resource.

A Seagull can feel overwhelmed if too many tasks are thrown at him or her willy-nilly. Because of the Seagull's great desire to have a "clean beach" at the end of the day, s/he will need permission to do small pieces of larger projects and have the sense of accomplishment and completion that s/he needs.

A Seagull needs frequent one-on-one attention. It does not have to be for a long period of time, but in order to anticipate your needs, s/he must be kept in the loop. Knowledge of your concerns, appointments, whereabouts, and needs are his or her source of power. Additionally, seeking his or her input on decisions, priorities, strategies for working with others underscores that you value your Seagull's expertise and confidentiality. Commands and directives are not generally effective with a Seagull.

BOTTOM LINE

In today's corporations, many of the Seagull's traditional support positions are being eliminated, with tasks formerly performed now being done by the professionals themselves or by a technical/administrative support pool. The pool does not permit the Seagull to have the close personal connection on which s/he thrives. The professionals and supervisors are using their time on tasks for which they were not trained, to which they are not inclined, and/or which are not the highest use of their time.

Many businesses have become report- and database-dependent, requiring complex, seemingly endless input by the service provider himself or herself. Therefore, the executive secretary or senior executive assistant positions are either broadening or going away. Examples include paralegals, veterinary technicians, etc. In the military or federal government where many senior commands are following the business model, the commanding officer or general is now a "program manager" and does not rate a senior secretary. So, the secretarial position must become that of an analyst of some sort. Another example from the medical field is that the doctor's right arm, his receptionist/nurse must increasingly be a nurse practitioner.

The Seagull is being robbed of prime feeding areas. Our corporate environments are becoming unbalanced. Hopefully, the corporate leaders will begin to understand that without Seagulls on their beach, they will soon find themselves overrun by Egrets, Cormorants, and Scurry Birds. These lack depth and breadth of knowledge and allegiance to the leader and the mission.

The Seagull has freed generations of leaders to soar, scout, scan, and yet have home base protected. S/he has held the kite string and enabled them to fly high, flexibly tethered, almost free.

They also serve who only stand and wait.

John Milton, *"Sonnet XV: On His Blindness,"* 1652.

Probable Seagulls that I know:

____

____

____

White Ibis

White Ibis

The White Ibis is the bird that you frequently see in advertisements for Sanibel Island. There are four or five of them, white with wonderful red bills and legs, feeding together at the edge of the waves. Their faces are actually red and their bills curve downward. They are about the same size as Snowy Egrets and often welcome one or two Egrets into their feeding flock, because they feed on different food. The White Ibises feed on small mollusks which they find by poking their bills into the sand at the wave line. Coquinas are favorites. Small bugs found in grassy areas are also on their menu. Their feathers are white with the exception of the black feathers at their wing tips, visible only in flight. In flight, their neck is out-stretched vs. Great Blue Herons and Snowy Egrets, who fly with a tucked neck.

Ibises tend to fly in strings, as opposed to formations, often flapping and gliding. On long migrations, they adopt the more efficient V-formation. They often soar in circles. When a flock rises up from the shoreline, it looks like a choreographed cloud of birds, graceful and quickly resettled after being disturbed or choosing a new feeding place.

Margin and shallow water feeders, their primary habitats are seashore edges, salt, brackish and fresh marshes, rice fields, and mangroves. One may find them in almost any American coastal environment in the summer season.

Their voice, rarely heard, is low and harsh. They typically mate for life.

FOWL FOLKLORE

Before the Egyptian god Thoth became a Wood Ibis (Stork) to escape Thyphon, he was the guide, teacher, ferryman, and conductor of souls to the afterworld. Joseph Campbell, in his book, *The Hero with a Thousand Faces,* describes this god: "Protective and dangerous, motherly and fatherly at the same time, this supernatural principle of guardianship and direction unites in itself all the ambiguities of the unconscious…." Campbell describes

these guardians as keeping the hero from passing from the known world with its boundaries into the land of the unknown. But, it is usually only by venturing into the unknown that the individual becomes heroic.

Another mythological figure is Demeter, the Greek goddess of grain (bountiful harvests) and the mother of the fair Persephone, who was kidnapped by Hades. Determined to get her child back, she withdrew from Mt. Olympus, with the result that nothing grew and no children were born. Finally, Zeus heeded her demands and enabled the return of Persephone, but only for two-thirds of each year... an interesting commentary on the need for a daughter to be separate from an all-powerful mother for portions of her life. Demeter represents generosity, nurturance, bountiful provisions, and persistence. Demeter's depression in her daughter's absence and her destructive behavior to gain her deliverance are part of the archetype.

OBSERVED FOWL BEHAVIORS

While they do feed singly in the mangrove swamps, wandering carefully over the roots of the trees, they are most often found in flocks of five to twenty on the beach in Sanibel. They arrive at sunrise and depart at sunset. They methodically move along the shoreline, going quite some distance in the course of the day on foot. They will move slowly past humans if no aggressive actions are perceived. They wade into mid-leg depth, while poking incessantly for mollusks in the wet sand. In tougher weather conditions, they will actually wade into low surf.

They clearly operate from the theory of plenty; every member eats solo, but maintains a close proximity to the next member of the flock. They move laterally as the waves come and go, so they look like a feathered white and red wave on the beach. They swoop and peck in the sand and walk at the same time; it is the equivalent of grazing vs. the herons and egrets who strike at their food from above. It is very rare to see two Ibises fuss over one another's catch.

On the mud flats at "Ding" Darling Nature Preserve at Sanibel Island, they look like the old-fashioned child's toy that has four

chickens whose heads bob as the ball on the string underneath swings to and fro. Their feeding movements are rhythmic, continuous, and precise. An Ibis always finds something worth his while.

There is abundance for all, and there is a willingness to collaborate with other species of about the same size who have a different diet. So, Snowy Egrets, who are often very territorial and feisty, will be welcomed into the flock as it moves along the shore line. The White Ibis is also happy feeding alongside a Great Blue Heron or among Roseate Spoonbills (though the Spoonbills enjoy the same small mollusks as well as their primary food, shrimp). The tiny Scurry Birds (Sanderlings) will often run in and out of the Ibis flock, undisturbed by the presence of the larger birds.

They keep their immature young, who are a mottled brown, with them in their feeding flock. They are also the only species this author has observed, other than the Scurry Birds, who keep their injured members in the flock and adjust their speed to accommodate the slower birds.

The White Ibis values group interaction, not only feeding together, but bathing and preening together after a morning's feeding. While other birds may flock together only during mating and rookery periods, the Ibis seeks out camaraderie throughout the day and at eventide.

OBSERVED HUMAN BEHAVIORS

The White Ibis sees life from a glass half full perspective. In an office setting, a White Ibis provides the reassuring "glue" that keeps diverse members of the staff working together without fussing. When there is a White Ibis in your workplace, there is a general sense of peace and well-being; his or her absence is sorely missed.

The key to the White Ibis's role is that s/he cares deeply about those around her, whether they are members of his or her biological or workplace family. S/he brings a concern for others' well-being that is selfless and nurturing. People find a reason to interact with a White Ibis on a daily basis. They know that it brings them balance and they sense that they are valued as individuals.

Senior Storks or other profiles can benefit by the presence of a senior White Ibis on their staff. When the Stork must issue edicts that feel impersonal or punitive, the White Ibis can help others settle down, unruffle their feathers, and take a more positive look at the fallout.

A White Ibis often plays roles that are tangential to mission; s/he holds positions in human resources, teaching, coaching, counseling, or nursing. S/he may also be a significant volunteer, coordinating others who support through service. His or her inclination to nurture vs. combat enables the Ibis to respond to people in trouble. However, his or her response is more like a hotline than an advocate; s/he listens and includes the injured person in the flock, but rarely goes to bat for the troubled individual or is forceful in getting the help needed. The Ibis is non-combative by nature. Only when one of her "own" is threatened, like Persephone, does the Ibis become enraged, and the strategy employed is usually not combative.

One side effect of this nurturing role is that a White Ibis can live vicariously through those whom s/he supports. In this sense, the White Ibis does not achieve on his or her own, but has an almost parasitic existence. As such, s/he is vulnerable to being taken for granted, to burnout and to pressures of all sorts. There is the potential for martyrdom for the failing Ibis.

The nurturing of a White Ibis may be manipulative or controlling, albeit couched in a silk glove. Love and service may be provided conditionally: "If you really love or appreciate me, you will do it my way." As with the god Thoth, his or her protection may be both nurturing and dangerous to those who must accept life's challenges to mature, to become heroic. S/he can resemble the parent figure who protects his or her young beyond the developmental stage of healthy separation. She can worry about the prospect of the "empty nest syndrome," which other birds do not experience. A supervisor, manager, or founding mother/father can do the same, expecting allegiance from his or her proteges and resenting any attempt to move on.

The Ibis does tend to see things as black or white, without much shading of grey, so s/he is not as imaginative and flexible in

response to challenges as some of the other birds. S/he plays exactly what is on the sheet of music; s/he follows rules, protocols, and standard operating procedures (SOPs), but does not fare well with changes. Inclined to a win-lose perspective, s/he is often on the losing end with more feisty, combative birds or more flexible, creative birds. Flight is his or her preferred defense. Whining or complaining is another response in the workplace.

In the same way that the White Ibis feeds incessantly from dawn to dusk, with a few social preening periods in between, the human Ibis works tirelessly. S/he can always find food (i.e., work to do) and others may stir some up for him or her. Because time seems to be an endless flow, s/he often does not learn the discipline of choosing and prioritizing commitments. S/he does not know how to say "no." A necessary life lesson is learning how to say, "Thank you for thinking of me for this task, but it does not fit my priorities at this time."

The Ibis can be a borderline introvert or extrovert. The key is that s/he is very focused on serving others. His or her leadership style is parental ("I know what is best for you"); and s/he often avoids leadership of peers, preferring to work with children or younger adults. An Ibis can often believe that s/he operates from a democratic approach (at home, in the classroom, or in the workplace), while the reality is a pseudo-democracy that can infuriate perceptive subordinates. His or her official power in the workplace is usually based on knowledge and, perhaps, seniority in a position; his or her unofficial power comes with a positive personality.

The Ibis is socially adept and reads people well, especially if they threaten those for whom s/he feels responsible. However, s/he does not read his or her own needs very well and may not take care of personal physical and emotional needs until s/he has no choice but to do so. Then, s/he feels like a failure for letting others down.

Again, due to great sensitivity to the nuances of others' body language and tone of voice, a White Ibis under stress may misinterpret the casual comments of others. The end result is that s/he can feel misunderstood and mishandled when s/he mis-reads

intent. This can be especially common as a White Ibis grows older and hard of hearing.

Manners and values are of great importance to the White Ibis. Genealogy may be of great interest to a White Ibis because of the support of s/he gives her own family. Jean Shinoda Bolen, M. D. in her book, *Goddesses in Every Woman: A New Psychology of Women,* notes that "within families, mothers and daughters who are all Demeter women may remain close for generations. These families have a decided matriarchal cast. And the women in the family know what is going on in the extended family, much more than the husbands do."

In order to be taken seriously and treated professionally, the White Ibis may need to disavow others of the perception that "anyone can do" what s/he does. This is the apple pie and motherhood curse in the workplace (when most of us would admit that parenting is one of the toughest challenges and few of us can make an apple pie!). Some ways to subliminally change others' perceptions are: to post one's credentials or certifications; to wear an official name badge or security pass visibly (if the work environment features seniors who do so); and to wear a uniform or quasi-uniform such as a minister's collar or a nurse's, doctor's, or dental hygienist's coat (they come in a variety of colors these days, so the Ibis could be less formal, but still get the point across). Another strategy is to do some of the bureaucratic tasks that are so hard to get completed by administrative staff in a timely fashion, such as travel or training orders, purchase requests, etc. The Ibis will learn the processes this way and not be at the mercy of others' priorities. S/he will demonstrate that s/he is a team player and is not dependent upon the whims of junior members of the organization.

VISIBLE SIGNS

The office of a White Ibis will be comfortable looking, with some "homey" or cabin touches. Floral bouquets and art work might grace a female Ibis's office, while a male might include a collection of pipes, photos of birds, animals, or his boat, and a plaid throw, pillow, or rug. The overall message is one of natural ease

and a delight in children, pets, and family. Colors will be soft and rich, and textures will add depth and warmth.

The White Ibis will wear comfortable, casual, classic clothing. A male Ibis will wear a sweater, a la Mr. Rogers, in his own neighborhood, but will have a sports jacket available if he must interface with his seniors. The female Ibis will also wear a cardigan sweater when she feels a more formal jacket is not required, and she will tend to select fuller skirts vs. a tailored look. Tweeds, mohair, paisley prints, and pastel colors will be favored by a female Ibis; the clan's plaid or corduroy will adorn the male.

HUMOR

The White Ibis delights in amusing stories about people, often those that poke fun at foibles or facades. S/he is not intentionally malicious in her humor, preferring to gently tweak others. She enjoys laughter. When you have a fair Ibis on your staff or in your group, you will be aware of smiles, delight in each other's company, and the strains of amusement in the halls. S/he brings out the best in many of the other birds.

POLITICS

Candidates are fortunate if they have White Ibis volunteers in their campaigns. They are loyal supporters and tireless workers, not looking for the limelight themselves, but relishing the success of the whole group. A wise political candidate compliments his or her Ibis volunteers frequently by saying, "I couldn't do it without you." (And s/he couldn't!)

FAIR FOWL

At his or her best, the White Ibis is a cheerleader and a coach for others. When others do not take his or her advice, the fair Ibis can gracefully say, "You're right. There is another approach to this issue or dilemma," and not feel as if s/he has been cast off. This mature Ibis can often be seen lobbying for legislation to correct ills for large groups of people. The motivation is service, not personal aggrandizement. As such, this fair fowl is a notable champion because his or her motives are pure.

The Fair Ibis often leads by default. S/he would not have chosen to play the leadership role, but events or the departure of a previous leader open a void that s/he recognizes must be filled. Wives of deceased Senators and Congressman have played this role. His or her tenure is usually a time of good will in the organization, a time of cohesion-building. When the organization needs to take risks and initiate new strategies, the Ibis will prefer to step down, having served a critical role in stabilizing the situation.

The Fair Ibis takes on tasks that at first seem very risky to him or her; s/he worries that s/he will not be able to produce as expected. But, motivated by wanting to support an admired superior, s/he finds himself or herself growing into the new role. Typically, s/he adds his or her own personal warmth to the delivery of the product or service, and is pleasantly surprised by others' high evaluation of his or her success. Laura Bush is a Fair Ibis. She has brought such credible compassion to the role that she plays, that many find her more appealing than the President.

The Fair Ibis is often a teacher or a coach who has supported whole generations of young people in town. S/he is revered by those for whom s/he was exactly the right support at a specific stage in their lives. Or, the White Ibis, in a Lord or Lady Bountiful role, will enable the start of a much-needed program. However, s/he will prefer to leave the daily work to an executive director, while s/he maintains a board or figurehead position.

A Fair Ibis in a workplace setting will be the person who always has a smile on his or her face, who arranges all of the decorations for the holidays, who coordinates the celebratory meals or birthday cakes, and generally thinks of the gracious things that will bring pleasure to a group that can get too focused on mission to think of themselves.

Gentle and not competitive, the White Ibis is often uncomfortable in settings like college reunions, where many of the other birds barely disguise their competitive achievements. S/he will be very careful about voicing opinions, not wanting to offend others who may believe differently.

Because s/he enmeshes her sense of who s/he is with those she loves and/or supports, s/he sometimes offends by not recognizing boundaries. The White Ibis can brag about his or her children or group to the point of becoming embarrassing, and can inadvertently divulge information that should not be shared with others.

FOUL FOWL

A generous person by any measure, the Ibis can be taken advantage of by others with more greedy profiles. The result can be martyrdom or rage. When a White Ibis believes that s/he is not getting appropriate recognition or support, s/he can become a dysfunctional worker. S/he will fail to give the courtesies to others that s/he believes are so important. S/he can become demanding vs. graciously appreciative. "What are you going to do for me?" becomes her leitmotif. Others in the workplace quickly cease the niceties that keep peace and well-being in the Ibis's world.

If a White Ibis finds himself or herself vying for the attention of someone important to him or her, s/he can feel very threatened by high-achieving birds. The White Ibis's primary gift is seemingly selfless service and devotion, which may not be as visibly attractive as other qualities and achievements. S/he may become petulant, pouting, and petty in response, because s/he is dependent upon those s/he serves. If s/he is not getting the resources or support s/he needs, s/he may become quite stubborn or dramatic in her demands. S/he may lash out at safe targets, who remind him or her of her loss; but s/he cannot afford to antagonize the important ones.

A White Ibis can be very vulnerable to Cormorants who maliciously manipulate the Ibis's tendency to welcome all comers until proven otherwise. The naivete of the White Ibis means that s/he will take a long time to recognize a Cormorant for what s/he is. When it finally happens, the Ibis's anger will be furious, like Demeter's.

Another tack is martyrdom. Studies over the last decade of those at risk for heart disease find that those who stoically hide negative emotions are four times more likely to die of heart attacks than those who deal with disappointments and anger openly. "These are the lady-like ones who take care of everybody but

themselves," according to Darla Vale of Rush-Presbyterian-St. Luke's Medical Center in Chicago.

STRATEGIES FOR INTERACTION WITH A SENIOR IBIS

Jean Shinoda Bolen, M. D., in her book, *Goddesses in Everywoman: A New Psychology of Women,* is illuminating about a "Demeter woman in leadership and founding mother positions." She says, "Conflicts arise within her and between her and those she supervises because she is a person with authority who sees herself and is seen by others as a nurturing figure. It is difficult for her to fire or confront an incompetent employee, for example, because she feels sorry for the person and guilty for causing pain. Moreover, employees expect her to look after them personally... and are resentful whenever she doesn't."

Dr. Bolen further notes that the Demeter woman "supervises closely and then always adds the final touches at the end. Whatever the daughter [or employee] does, the mother [supervisor] gives her the message that 'It's not good enough' and 'You need me to do it right.'... which stifles originality and self-confidence in her 'child' and increases her own workload." This scenario can quickly lead to burnout and/or martyrdom.

A strategy for healthy interaction with the senior White Ibis, then, is to ask for his or her guidance before starting a project. The first time through, check in with him or her as you progress, so that small tweaks give him or her a sense of ownership and control and you will not waste energy and your own self esteem by getting too far ahead on your own. The next time, you can diminish the number of mid-points, once s/he has confidence in your work.

A senior Ibis is particularly well-suited to supporting employees who are new to the workplace and/or profession, as well as those who have special needs due to disabilities. A senior White Ibis also can be a very effective leader for volunteers because s/he provides the warm structure that allows them to feel productive and appreciated — usually their only reward.

Because s/he so values the personal services aspects of his or her job, s/he can benefit from assistants or team members who prefer the non-personal tasks such as scheduling appointments, maintaining records, creating newsletters or flyers, etc.

If you work for a White Ibis, be careful about what you share that is personal, whether delight or distress, because s/he will feel compelled to advise you about it and will be likely to share it with others — not as gossip, but just in the sense that you are "part of the family." It could be very embarrassing, even though the disclosure was not meant to be malicious.

The White Ibis will need your constructive input delivered in a way that s/he does not perceive as personally critical. One approach is to comment positively on an aspect of the project and then suggest ways to enhance the outcomes even further. Your body language needs to be warm, not confrontational. This gives room for him or her to feel complimented and supported, and then make some twist of his own so that any changes feel like his idea.

STRATEGIES FOR INTERACTION WITH A JUNIOR IBIS

Often a White Ibis has been a mother who focused all of her energies on her family. She had an extensive membership in a variety of organizations, though typically was not the leader. When her children leave home or when divorce or widowhood precipitates her movement into the workplace or a significant volunteer role, she is ill-equipped to thrive on her own initiative. A wise supervisor can assist her acculturation by finding ways in which she can gain some of the affection and admiration that she craves, but also produce structured accomplishments. She needs to have a bridge to self-achievement in order to balance her propensity for group immersion (and the resulting lack of self-esteem).

Another common Ibis prototype is a man or woman who has completed a career in a field that did not provide much outlet for his or her nurturing, coaching, or pastoral inclinations. So, for the second career, s/he chooses a setting in which personal interactions

are more fulfilling. S/he has significant life experience, but will need mentoring in appropriate interface and protocols to protect him or her and the recipients of service from enmeshment. Because s/he likes dependents, s/he is tempted to provide for clients or students instead of teaching them how to provide for themselves, which is the ideal long-term goal.

In staff meetings, s/he looks like a non-participant by design. S/he is listening intently, but rarely feels compelled to add his or her perspective publicly (though s/he may do so when queried gently in a one-on-one conversation).

His or her identity is entwined in the whole group. It is important for a manager or supervisor to evaluate the White Ibis separately and note particular achievements or areas needing improvement. S/he will be particularly vulnerable to broad-sweeping comments or negative critiques by management, because s/he does not easily think of her own role or responsibility for whatever the issue is.

Because the Ibis is so responsive to others, s/he may not take or protect the time and energy necessary to produce the quality product or service of which s/he is capable. A supportive manager can help him or her schedule blocks of uninterrupted time, knowing that the Ibis will still need social interactive time to re-energize.

There is a tendency for a program or a project to become the "baby" of a White Ibis. While a supervisor may cherish the commitment made, it may become problematic if the supervisor decides that the program or project must be diminished for any number of reasons. The Ibis has identified so thoroughly with it that s/he cannot see the program's relationship to a larger whole and feels personally attacked. A key strategy for the supervisor is to talk openly with the Ibis on positive and negative issues, and attempt to delineate the difference between the personal and the programmatic when downsizing or restructuring must occur. If the supervisor can involve the Ibis in the downsizing efforts, then the Ibis retains a sense of control and value to the organization. However, there are times when this collaboration is not possible, due to contractual, union, or other legal constraints. In cases like these, the supervisor must be prepared to have a very hurt Ibis on his or her hands. Passive aggression will be a likely result, as the Ibis does not relish confrontation.

BOTTOM LINE

White Ibises can be truly generous, or may be selfishly giving in expectation of control and obedience. In order to be their best selves, they need some introspection and they need to value themselves — a seemingly stingy posture to many.

> *Girls and women thrive in relationships; for women the apex of development is to weave themselves zestfully into a web of strong relationships that they experience as empowering, activating, honest, and close.*
>
> Jean Baker Miller,
> Stone Center for Research on Women, Wellesley, MA

> *When threatened with a major illness, a White Ibis responded: "I have never prayed for myself — for everyone else, but never for me." When advised that maybe she had more than earned the right to pray for herself, she looked embarrassed and anguished, but also hopeful.*

> *At D. C. General Hospital (the city's only public hospital): On the cancer ward, a nurse made her rounds with dollar bills stuffed in her pockets. She never knew, she explained, when a patient might need bus fare home.*
>
> The DAY, New London, CT (May 7, 2001)

Probable Ibises that I know:

____

____

____

____

Roseate Spoonbill

Roseate Spoonbill

Roseate Spoonbills are one of the favorites with birders because they are so colorful. John James Audubon discovered these birds in 1832; a century later, they were so endangered that the Audubon Society's first director of research, Robert Porter Allen, could not kill one for study. Instead, he lived among them and changed the way research is done.

At 32", they are a bright pink wading bird with a flat bill that they sweep rapidly from side to side. A combination of diet (accumulated iodine) and age produces their conch-shell pink coloration, plus red on their shoulders and orange on their tail. Much like the stork, they have a naked, gray head. The young are white until they are about three years old.

Spoonbills fly in V-formation or lines, often gliding between a series of wing strokes. Unlike the Herons and Egrets, they fly with their neck stretched out.

Their primary food is killifish, minnows, and shrimp in shallow waters: coastal marshes, mud flats, lagoons, and mangrove keys. Other foods include snails, aquatic insects, and aquatic plants; they are opportunistic eaters. They essentially catch floating prey by waving their open bill from side to side; their bill snaps shut when a floating or swimming object touches the inside of their bill tip.

FOWL FOLKLORE

Persephone, the daughter of Demeter and Zeus, was snatched by Hades and taken to the underworld as his bride. In her beautiful maiden persona, she was symbolized by the narcissus, as well as corn and grain. When she was rescued, she was queried by Demeter if she had eaten anything while in the underworld. Persephone lied to her mother, saying that Hades had forced her to do so. As a result of her foolish consumption of pomegranate seeds, she was required to spend one-third of each year with Hades. She is comfortable in Pandemonium (the chaotic home of the demons in Milton's "Paradise Lost").

The Temptress in literature is the seductive and beautiful woman who brings about the destruction of those she ensnares .... often by deflecting them from their journey or quest. An interesting twist on this theme is found in the story of Dulcinea (Aldonza), Don Quixote's "lady." She tells a tale of woe and whoredom, but in the end, comes to Don Quixote because she needs to believe in something better. So, if a Spoonbill can align himself or herself with a very ennobling vision(ary), s/he can rise above his or her tawdry propensities.

In children's literature, Eyeore is the classic sad sack and pessimist.

OBSERVED FOWL BEHAVIORS

The Spoonbill is a tall wader in the shallow bays at "Ding" Darling National Refuge on Sanibel Island. His rhythmic swinging of his bill through the water enables him to scoop up all sorts of edibles, much like a pink-legged vacuum cleaner. He filters his intake, then shakes his bill to sluice out the excess. Spoonbills feed in groups or singly, like the White Ibis, and there is no squabbling for territory such as one sees with Snowy Egrets.

He rarely makes a noise, even though he is in close proximity to many others of his flock and others who look for different food. One will see Snowy Egrets (spaced throughout the flock) and lots of Scurry Birds on the mud-flats. If Storks appear, the Spoonbill flock will find another feeding area.

The Spoonbill is primarily a nocturnal feeder because that is when dense shoals of fish pour into the shallows. Also, a ranger at "Ding" Darling National Refuge explained that the Spoonbill "can't stand the heat." So, he roosts all day in the trees and feeds at dawn and at sunset.

OBSERVED HUMAN BEHAVIORS

When one first encounters a Spoonbill, one is often enchanted by his or her attractiveness and engaging, almost child-like interaction. S/he can be delightfully spontaneous, gregarious, playful, and generous. However, this high energy profile takes a

great deal of effort for the Spoonbill, and down time will be required. His or her energy comes from focusing on personal needs as opposed to the needs of others. The Spoonbill's egocentric behavior is not malicious; it is simply self-centered and s/he is usually oblivious to others' sensitivities. Unable to focus on others' conversations very long, s/he will interject with something totally irrelevant to the discussion in progress, but highly relevant to the Spoonbill. This occurs because the Spoonbill views the universe as finite vs. infinite: "if attention is being paid to someone else, it is not being paid to me" vs. "there is enough attention to go around."

The Roseate Spoonbill sees things as black or gray and operates from a lose - lose philosophy. Nothing good happens for long from the Spoonbill's perspective because his or her glass is half empty. When others achieve or receive something that the Spoonbill envies, s/he sees them as lucky and himself or herself as unlucky. The concepts of earning and deserving are foreign paradigms.

Much like the bird, the human Spoonbill gets "pinker" the more s/he feeds on self-attention or self-pity. While some birds need to have some confrontation in their diet before noon, or they go looking for it, the Spoonbill needs some compliment or put down, perceived loss, or unfavorable comparison to others before noon.

Ironically, if some dysfunctional Spoonbills find themselves at risk of high self-esteem, they will behave in such a way as to cause others to react negatively, thus returning them to their pink condition. A failing Spoonbill who has just been complimented on her performance will turn around and blatantly refuse to do a routine task or tell her superior that it will be days before s/he could get to it. While many adolescents go through a stage that bears similarity to this seemingly random collaboration and then rebellion, the Roseate Spoonbill follows this behavior pattern for a lifetime.

The Spoonbill is usually an introvert who is so internally focused on his or her own needs that s/he wearies others who do not want a "dependent." S/he is happiest when the center of attention (for good or ill). A real "attention-sponge," s/he resents others' happiness and often will attempt to make them feel guilty

for it. Attention for the Spoonbill is a debit card; there is little banked for another day.

His or her finances also resemble this pattern, with an inability to control finances (and a parallel sense that there is no rush or need to repay any loans from friends). There is a sense of entitlement. Because a Spoonbill is so riveted by the present, s/he rarely plans for a rainy day or retirement. Savings and investments seem boring, and a large credit card debt is a badge of honor.

A shallow water feeder who vacuums methodically for morsels, the Spoonbill enjoys the company of other Spoonbills and other species who stir up the area. White Ibises and Spoonbills are often found together in a human setting; each provides what the other wants. The Ibis wants dependents to care for and support; the Spoonbill wants the attention and is happy to be dependent.

Gossip of all sorts is the equivalent of shrimp. The Spoonbill finds its food by feel vs. seeing it and strategizing how to catch it. S/he is governed by emotions vs. analysis of facts and situations. More is better, especially if it is more than someone else has; there is little sense of self-discipline. S/he is subject to sensory overload because s/he takes in enormous quantities of experience (or work). If his or her filtering mechanism is not judicious, s/he can easily become overwhelmed.

S/he ensnares others because s/he is persistent and flirtatious in having his or her own needs met. Passionate, pandemoniac, and flamboyant, the Spoonbill finds that life is not as real as novels, movies, soap operas, and dreams. These romanticized worlds appeal to the Spoonbill. As s/he looks at his or her own life, s/he will embellish events to make them more extraordinary. Telling his or her story is very important; relating the number and details of romantic interests or conquests s/he has had allows the pink factor to rise.

Exercise provides a way to look like the perceived ideal (which is usually a great deal thinner or more muscular than a healthy norm). Exercise can also be a way to work off psychic damage from powerful others. Anorexia and bulimia can plague the Spoonbill, but not due to self-control such as the Egret's; rather, as a reaction to others.

The Spoonbill is a consumer and measures self-worth by what s/he has. S/he expects that others will provide for him or her, but forgets quickly any benefits received and soon whines, "What have you done for me today?" S/he expects the best equipment, the best benefits package, etc. without commensurate personal achievement. S/he does not see it as unreasonable that as a junior member of a staff, s/he wants the newest computer or software in the office. Often, the Spoonbill has a very creative rationale for desired, but unearned, perks.

The Roseate Spoonbill can be quite creative and socially adept when "up;" but the mood swings are frequent and unpredictable. From his or her perspective, others are to blame for all of the bad things that happen to him or her. The Spoonbill has a responsibility deficit. When actually confronted with the facts regarding events, s/he will have invented a story line that s/he really believes (though others may see as a lie). Eye contact will be aggressively direct or averted.

The Spoonbill will spend a great deal of time combatting the evils in his or her world — usually on a personal level. Unlike Don Quixote who tilted at symbolic windmills for the benefit of mankind, the Spoonbill tilts at those who have slighted him or her, or worse yet, have deprived him or her of what s/he believes is due. The Spoonbill can be easily manipulated by others for their own malevolent purposes, so s/he is often found in the company of Cormorants.

Frequently there is a history of significant weight as a child or some other challenge that made him or her feel unattractive. The Spoonbill overcomes that deficiency physically with a great deal of effort, such as the prolonged wearing of braces, but the emotional scars remain. There are parallels to the story of the ugly duckling. When one's exterior changes, but the interior terrain does not, there is a sense of not deserving others' praise or attentions. There is also a tendency to accept the first proposal (whether that be marital or employment-related), because the glass is half empty ... the Spoonbill might not have another chance. This precipitous decision-making resembles those with Attention Deficit Disorder.

While very changeable, the Spoonbill really is not flexible, because flexibility requires constants from which to vary. The Spoonbill is in a flux, not in a flex mode.

Even though s/he is keenly aware of his or her own superficial feelings, the Spoonbill has little intra-personal intelligence. However, s/he can read others well (inter-personal intelligence) when the motivation is to determine how and if they will support him or her. The interest in others' moods is not based on altruism. If others are sad, the Spoonbill may feel sympathy (because experiencing the sadness feeds him or her); s/he is less able to feel empathy, a distanced understanding (and analysis) of others' needs. A Spoonbill counselor should insist upon boundaries with clients, because s/he does not have enough emotional/psychic energy to share their woes as well as his or her own.

VISIBLE SIGNS

A Spoonbill often creates a non-traditional office space by using unusual paint colors, bold or "romantic" artwork and decor, and furniture arrangement which s/he changes frequently, as s/he is always looking to upgrade.

Order in one's workplace is not required by a Spoonbill. Disarray is not a deterrent, nor an embarrassment; s/he does not feel obliged to remove his or her personal belongings from commonly shared spaces like lunch rooms or changing rooms. His or her office may look like a cyclone has hit it. While that may be all right in his or her living accommodations, it rarely meets the standards for a workplace, especially a workplace in which customers or clients are served.

A female Spoonbill is fond of perfume, scented candles, and makeup. Both male and female Spoonbills can spend a lot of time and money on their wardrobes. A Spoonbill is making a statement with his or her apparel, which usually is quite colorful. Some will use their outward appearance (body piercing and brightly dyed hair) as a rebellious statement. Much like adolescents, they are reacting against others, as opposed to being truly self-directed. A Spoonbill can be a perpetual adolescent.

HUMOR

The Spoonbill will be intrigued by caricatures. These exaggerated profiles are right on for this person who is so very interested in personal and performers' idiosyncracies.

This is the artist who paints endless variations on a theme: himself (or herself). S/he often has a self-deprecating style of humor, which prompts others to say, "Oh, no, you're too hard on yourself." S/he will also relish off-color remarks, especially if they relate to his or her own body parts.

The Spoonbill enjoys the company of a Pelican because of the infectious high that the Pelican generates as s/he gets rolling.

POLITICS

If involvement in politics will bring attention, count on the Spoonbill. Spoonbills like the ups and downs, the drama, of politics. They are not intrigued with the homework portion of being a good representative, however. The publicity, parties, perks, and performances are the fun portions. Spoonbills are politicians of excess, taking every trip available. They are vulnerable to manipulation by lobbyists and members of their own staffs who have their own agendas.

Another form of political involvement by Spoonbills is the public hearing or call-in talk show. These offer a forum for visibility that feeds their need to feel important. Many of these settings resemble soap operas.

FAIR FOWL

The Fair Spoonbill is "up" most of the time. Because s/he has remarkable intra-personal intelligence (i.e., s/he really understands himself or herself) and s/he recognizes the needs of others, s/he can be a skillful and empathic counselor. S/he understands the desirability of teaching others "how to fish;" the dysfunctional Spoonbill will insist on doing the fishing, because that is how s/he gets the attention and gratitude s/he craves.

Because s/he has had to deal with his or her own personal "demons," and has done so with significant psychic energy and focus, s/he can be a role model for others. S/he is the columnist, actor, or musician who has learned how to use personal experiences on a higher level, using the insights and emotions to enhance his or her art.

A Fair Spoonbill can be an incredible creative, thoroughly capable of a one-man/woman show. It is harder for him or her to collaborate unless s/he is given a discrete arena of his own for which s/he is responsible and in which s/he has the freedom to experiment. Bureaucracies are not conducive environments for a fair Spoonbill.

There is a danger that this high-functioning Spoonbill can be manipulated by others. S/he must be on guard for those who know that his or her high functioning behavior requires a great deal of energy. Those who want to take her down a peg or two can simply fail to support her project or program. They can second-guess her request for greater personal security in the workplace, belittling a very real challenge to her safety by an enraged client or co-worker.

The Fair Spoonbill functions much like a Great Blue Heron. Therefore, the backlash that appears "from nowhere" is unleashed by those who see his or her performance as challenging to them. In this case, "It is about <u>them</u>" and not about him or her. This backlash to gifted, achieving individuals can be recognized by its emotional fervor and the persistence of the perpetrators.

FOUL FOWL

The Foul Spoonbill is "down" most of the time. The mood swings are problematic in terms of completion of work and interface with customers and co-workers. If the Spoonbill spirals downward, s/he may have manic-depressive and bi-polar tendencies. Caring family members and friends should help a dysfunctional Spoonbill get the kind of mental health therapy that s/he needs in order to function safely. Otherwise, at-risk behavior multiplies. Suicidal tendencies are present. Abuse of drugs or alcohol is a common "solution" to a Foul Spoonbill's distress and

dilemmas, which are always greater than anyone else experiences, s/he thinks.

The failing Spoonbill has highs of confidence and pits of despair — but sees someone else in shining armor as her way out. The male version of this attaches himself to a strong woman who finds him appealing because of his neediness and "little boyness."

The failing Spoonbill uses his or her powerlessness, helplessness, martyrdom, and even guilt to control others. "Poor me..." There is a pathos in most of his or her stories. S/he will have tales of trying to be a good whistle-blower or advocate, only to be treated badly by superiors. S/he will have the reputation of having been a trouble-maker wherever s/he goes. (Other birds are accused of this in order to diminish their successes; however, the Spoonbill probably has made a mountain out of a mole hill.) S/he seems desperate for attention.

When a Spoonbill attempts to get one to intercede or be a go-between with a third party, it is important to insist that the Spoonbill go directly. Manipulation of this sort is called triangulation, and is not a healthy role in which to participate. One finds this in families as well as workplaces.

The failing Spoonbill is pressed to impress. This bird is so focused on himself or herself, that all conversation is manipulated to come back to him or her and s/he talks non-stop. Exaggeration and embellishment are constant; it is hard for others to sort out what is true in this soap opera kind of conversation (and it is hard for the Spoonbill to remember reality as well, because s/he has created so many versions of his or her stories that they have become real).

The Foul Spoonbill is not good at reading others' moods or reactions because s/he is so focused on herself. When threatened, however, s/he becomes more observant, almost obsessed with catching his or her challengers in a "gotcha" mode.

Flirtatious and coy behavior becomes so habitual that s/he is surprised when some respond as if s/he intended the pattern. S/he is clueless to what s/he might have done to provoke the sexually-loaded response. The "come-on" is always turned on.

STRATEGIES FOR INTERACTION WITH A SENIOR SPOONBILL

It is rare to find a Spoonbill in a senior position. Usually, seniority is based on one's contributions to the workplace, and Spoonbills are too self-focused to perform well consistently enough to earn major promotions.

However, in the instance in which one finds oneself working for a Spoonbill, the strategy is to provide the attention that the Spoonbill needs in order to get pink enough to focus on the tasks at hand. Expect volatile behavior. One day you will be God's gift to the department and the next you will receive a tongue lashing for failing to serve the Spoonbill in a previously unsought way. If you can be the solid underpinning that the Spoonbill needs and can let the dramatics roll off your back, you might be able to demonstrate your worth to the workplace and get promoted by someone senior to the Spoonbill. S/he will not be inclined to let you go if you are meeting his or her needs.

A senior Spoonbill may expect friendship from her favorite reports. This over-stepping of boundaries can be dangerous for the proteges. According to researcher Susanna Rose, women, in particular, often define friendship by one's willingness to share fears and disappointments (vs. achievements). "Thus, in friendships women emphasize their vulnerabilities rather than their skills, their helplessness rather than their power." This is not a professional stance for the junior member of the dyad. The expectation by the senior Spoonbill that one's friend/protégé would always be supportive can be problematic when occasions call for honest disagreement and even competition.

The Navy training manual, "Human Behavior," succinctly sums up the dilemma of a senior Spoonbill: "A leader who is socially insecure has difficulty because the dedication is to gain affection, not to fulfill the mission. Sooner or later, the position [sic - situation] will arrive where loyalty to superiors will demand action that will not be popular with the people."

STRATEGIES FOR INTERACTION WITH A JUNIOR SPOONBILL

Not given to a sense of personal responsibility, the Spoonbill generally will not fare well in a collegial environment. The more traditional hierarchical structure will help to keep a Spoonbill in bounds. S/he would prefer to feed (work) whenever it suits him or her, and would prefer to have fewer rules and protocols, but s/he needs these structures to ensure equity in the work setting. S/he would take advantage of a more flexible environment.

The Spoonbill will require a great deal of attention and support from his or her supervisor in order to feel sufficiently appreciated and focused to do a good job. Ironically, this attention sponge will not reciprocate when the senior needs support.

Often very creative, the Spoonbill feels that s/he has produced something remarkable. S/he may have, but typically, someone else has had to do significant editing, formatting, or shaping in order to bring it to fruition. It is important to give credit to the Spoonbill where due, but also help him or her appreciate others' roles in completion of the project.

Very adept at giving reasons why s/he cannot do something, s/he will regale a supervisor with how the world is stacked against him or her. "They don't sell appropriate clothes in my size." "I can't drive, because I had an accident four years ago and now don't go beyond a one mile radius." "I can't take courses I'd like to take (or join service organizations, etc.) because I would feel guilty asking someone for a ride." One response is a simple nod or "Oh?" — but no attempt to fix anything for the Spoonbill. Another response for a supervisor is to structure his or her involvement in an event or project, particularly in some visible task; the result may be a Spoonbill beaming with pleasure.

S/he can cry at the drop of a hat. One strategy that workplace counselors have found effective with Spoonbills is to have them write down issues and explore them analytically. Studies demonstrate that writing produces positive chemical changes in the body; so, journaling helps diminish stress and helps the psycho-neurological immune system.

There will be times when a supervisor must get the attention of the Spoonbill who selfishly believes the world turns around him or her. "It's not about you" will be the caution. During periods when a workplace is experiencing downsizing or reorganization of tasks and roles, the Spoonbill will see himself or herself as the pivotal employee. The reality of his or her position in the pecking order may need to be made clear. Because of the Spoonbill's inability to be concerned about the department as a whole or its mission, s/he may find that his or her pessimism becomes self-fulfilling. It may be an opportune time to rid the workplace of someone who requires an inordinate amount of support to function well or even moderately.

A caution for supervision when the Spoonbill is of the opposite sex: because of the Spoonbill's tendency to be flirtatious, often without realizing that that is his or her affect, it would be wise to meet in an environment with others around. A closed door is not a good idea.

BOTTOM LINE

It would be wonderful if the human Spoonbills had such an outwardly readable meter of their well-being as the birds' pink coloration .... if one could readily determine that his or her Spoonbill is really in need of "TLC" today or is full of good strokes already. Does s/he already have his or her share (actually more than her fair share) of perks or strokes? Or is s/he anemic?

> *When a jealous friend causes you pain, it's a good idea to remember that termination of the friendship is an option... Jealous people aren't capable of true friendship, and jealousy and intimacy make a dangerous combination... An intimate relationship tainted by jealousy can be incredibly destructive.*
>
> Harriet Braiker, Ph.D., *Lear's* (Sept. 1989)

Probable Spoonbills that I know:

___

___

___

___

Brown Pelican

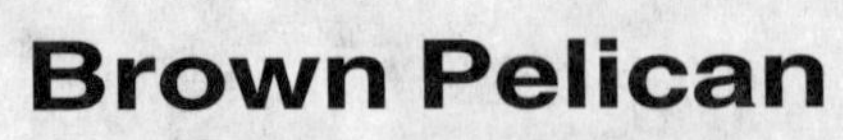

Brown Pelican

The Brown Pelican is the fellow who sits on top of a mooring pole at the ferry landing or at the marina. At one time an endangered species, the Brown Pelican is again soaring alongside you as you drive the causeway to Sanibel Island. He is your escort to all of the adventure that lies ahead.

A big bird, he is 50 inches tall and has a 6 1/2 foot wing spread. Though he has a broad chest, he maneuvers adeptly and has remarkable buoyancy. As an adult, he is a grey-brown with white around his head and neck. His most notable features are his enormous bill and expandable throat pouch with which he scoops up fish. Unlike other Pelicans, the Brown Pelican dives head-first into the water to catch his prey. (White Pelicans steam like a convoy, with Cormorants in their midst stirring up their prey; they scoop up fish from a floating position.)

Brown Pelicans are efficient fliers, alternating powerful flaps with short glides (while they peruse the landscape or seascape below). Flying in lines with their head drawn back to their shoulders, they are so close to the water that they almost touch the waves with their wing tips. When ready to feed, however, they get enough height (about 30 feet) to plunge head-first into the water (and then bob on the surface as they consume their catch or shake their head, having missed their intended meal). Their favorite food is small fish, such as menhaden and anchovies; they do also eat crabs and shrimp. Their pouch can hold about seven pounds of fish and water. They are often seen stretching their neck and pouch in order to keep the skin flexible.

Found in salt water bays and oceans and their adjacent beaches, the Pelican rests on posts, boats, and trees. The adult's low croak is rarely heard, but his progeny are very noisy. They mate for one season only.

FOWL FOLKLORE

While Bacchus (or Dionysus), the son of Zeus and a mortal, would be an extreme representation of the Pelican, he was the god

of wine and appetites of all kinds. In one of the ironic stories in mythology, King Midas was allowed to make a wish of Bacchus, having returned one of his company who had strayed. Midas wished that all that he touched would turn to gold. Bacchus granted his wish, only to have him plead for a reversal when he could not eat or drink. Bacchus was not interested in the accumulation of wealth, only in the enjoyment of nature's bounty.

Bacchus/Dionysus represented the dichotomy of kindness, the peace and beauty of nature, and food and drink, but also savage brutality. These two ends of the spectrum will show up in the Fair Pelican and the Cormorant (because bestiality and brutality belong to the domain of evil in our collection of behavior patterns).

Peter Pan never wants to grow up; he wants to live without responsibilities and schedules in Never Never Land. When Wendy (a White Ibis) comes into his life, he is sorely tempted by her offer to return with her and her brothers to her traditional life. But, he perseveres in his life of adventure. So, Peter is not in a stage of maturation; he has a core behavior pattern that will still be with him when he begins to grey.

Magnum P.I. gives us another view of an adult Pelican. Surrounded by friends to whom he is loyal in his own way, he frustrates his Snowy Egret, Higgins, by his cavalier sense of time and use of others' possessions.

OBSERVED FOWL BEHAVIORS

The head-first dive of the Brown Pelican is really an amazing sight and sound. His swift descent into the water can be heard at some distance. He plunges right in and emerges with a fish in his bill in one fell swoop. He then tips his head to one side to empty his pouch of water and points his bill upward in order to swallow his catch. Sometimes he will dive right into shallow water; evidently his buoyancy saves him from disaster.

If he is satisfied, he will just sit and bob on the water, fluffing his feathers with a seemingly satisfied preen. If still hungry, he is back in the air again. His great webbed feet help him paddle on the surface and are critical to his takeoff. When there are large

schools of menhaden near the beach at Sanibel, there are endless Pelicans swooping, plunging, and swooping again. The Terns and Laughing Gulls, who have learned that they can snatch fish from the Pelicans' bills, will be noisily announcing the feeding frenzy.

The Brown Pelican appears comical, but really is quite strategic and deft. When traveling, as opposed to feeding, he flies low to the water, getting a lift from his proximity to the water's surface. When feeding, he flies without casting a shadow that would alert his prey. On a feeding patrol, Pelicans often fly wing tip to wing tip, like a feathered airborne net.

Exceedingly gregarious, he may also be seen alone or in small groups. He accommodates the Terns and Laughing Gulls who literally land on his head as he emerges from a dive. His perspective is one of infinite abundance.

Pelicans are really loose in terms of leadership; they function more as a collaborative. If they are flying and searching for food, whoever ends up in front takes the lead momentarily. On the next swoop, another bird will lead.

They are extremely lightweight, being able to sit at the very end of an Australian pine bough and barely bend it. Their visual mass makes viewers do a double take.

OBSERVED HUMAN BEHAVIORS

The Nature Company Catalog (Spring 2000) featured an American White Pelican with his feet in a dance, his wings pulled back, and his pouch open as if he is laughing. It is the consummate pelican having a good time.

The Pelican is an easy-going, adventure-loving optimist who sees mostly white (on the black/grey/white continuum). His or her glass is not only full, it is overflowing. S/he operates from a win-win position. An extrovert, s/he is the most charming of the behavior patterns.

Much like the bird, the Pelican seems to coast through life, exerting energy only when necessary. However, it is easy to forget that prior to the glide, there has been a period of vigorous flapping.

The Pelican is often very creative, and may exert a great deal more effort than most observers (including bosses) give credit for. As a result, his or her achievements may seem effortless to others or s/he may seem to be goofing off and not working. S/he is perceived by many as superficial, but that perception does not take into consideration the risks taken by this aerial diver, nor the flashes of brilliance of which s/he is capable.

S/he possesses incredible enthusiasm for the overall scope of a project, but dislikes the details of execution. S/he can design a beautiful garden, but the nitty-gritty of planting each and every stem of pachysandra saps his or her energy. Wise leaders place a Pelican in position to soar as much as possible.

S/he brings fun to any group of which s/he is a part. S/he loves a game or sport and enjoys doing well. However, s/he is not "keeping score" all of the time. Periodically, the Pelican will outplay a Stork, much to the Stork's consternation and everyone else's surprise.

Spontaneous, s/he is always ready to stop whatever s/he is doing and join a group headed off on a foray of some sort. Not given to long-range planning, s/he is more opportunistic in approach. Sometimes a Pelican has a mental disconnect between his or her current status and a dream job or accomplishment. S/he sees just the "now" and the "then," but has little concept of the necessary steps in between. Underachieving gifted students are often plagued with this same difficulty. The other factor is that a Pelican generally is not ambitious in a competitive sense. S/he makes a good solo performer and sometimes a team performer, but not a leader. S/he definitely should not be expected to provide structure and supervision for other employees.

S/he often has a short attention span and a high distractibility factor. Part of this is true because s/he quickly gets 60 percent of the value of an experience and is ready to move on to new terrain. Other species want to have consumed every morsel available before moving on, because their paradigm is that of a finite supply vs. abundance. For example, the Pelican is one of the consummate television channel changers. Give a Pelican a remote control and s/he is in heaven, but others may be driven to distraction!

The Pelican has great eye contact because s/he is measuring his or her impact on listeners. Communication is generally one-way with a Pelican. If s/he is asked a question or is given an opening in a conversation, s/he will take it and run.

There is no dearth of IQ. Others may diminish the results a Pelican gets because s/he makes it look so effortless. Therefore, a wise Pelican will be specific about his commitment to a particular project and check in with his supervisor periodically to update completed portions and the efforts expended.

An improvisation artist, s/he is agreeable to try any number of approaches. Sometimes other species see this as lacking a foundation or core. A Pelican finds change a delight and sees continuity boring. "Let's try it!" is the Pelican's motto. "What have we got to lose?" Because s/he is alert to so much around him or her and is willing to risk, the Pelican joins the Great Blue Heron in terms of significant breakthroughs. The Great Blue Heron is more methodical in approach, but the Pelican's sheer volume of trial and error attempts (like his plunges for food) result in a high level of success.

Much more keenly sensory than many of the other birds, the Pelican loves color, unusual tastes and smells, music and a variety of people. S/he can be overwhelmed by sensory experiences. This is the architect who feels faint upon encountering a beautifully built environment. This is the artist who must capture what s/he has just seen or heard, before the fleeting vision or sound evaporates. The Pelican is keenly aware of the ephemeral. S/he is sensitive to the aesthetics of everything encountered, to the extent that s/he cannot work productively in surroundings that do not please him or her.

Because s/he is so sensory-oriented, there may be a dearth of abstract comprehension. A Pelican may be very literal and concrete. Others may think that s/he is purposely being obtuse, but the larger picture may not be as intriguing to the Pelican as the fleeting moment.

The Pelican is a generous bird and surrounds himself or herself with a wide array of friends (whose diversity is remarkable by any

measure). His or her favorite tools are the telephone and e-mail, because s/he is always ready to "reach out and touch" everyone. This happy-go-lucky individual does experience the fullness of life with an almost child-like wonder. S/he may invest more in friendships than others are able to return.

S/he is a wonderful member of an organization undergoing significant change, because s/he has a social and mental flexibility that allows him or her to see the positive possibilities ahead. However, s/he will not want to produce on time, in detail, or with long/short-range plans. Also, s/he does not like to provide the supervision /support for other employees.

VISIBLE SIGNS

The Pelican must have an attractive place in which to work. Otherwise, s/he will be distracted by things that displease him or her. A wise supervisor allows a Pelican to paint and decorate his or her own space. S/he often enjoys music while s/he works, but can use a head set to eliminate distractions for others — and, vice versa, other noises in the workplace will not distract him. Because s/he is intrigued by all activity, it would be wise to have his or her desk facing away from an entrance.

A bulletin board or its equivalent is very appealing to a Pelican, as s/he likes to have an ever-changing display of ideas that might trigger solutions for his or her current projects.

Clothing is commentary for a Pelican, but s/he does not generally follow business dress fashions. More casual and functional, albeit colorful, would best describe a Pelican's wardrobe. S/he likes costumes; cowboy boots or hat, a fisherman's vest, or a patchwork quilt jacket may signal a Pelican. Jeans are another favorite, because the Pelican likes sports, nature, and non-traditional activities and wants to be ready to participate on the spur of the moment.

HUMOR

While the Pelican shares an enjoyment of caricatures with the Spoonbill, s/he is more tuned into the foibles, the words and

gestures, linguistic mannerisms, and thought patterns than the Spoonbill. His or her mimicry is both gently humorous and highly perceptive; it is at a level that the Spoonbill (and many others) can enjoy, but not produce. Once on a role roll, the Pelican can pursue the monologue (or dialogue with a partner) beyond the endurance of most listeners, because either their cheeks ache from laughing or the inherent joke has gone stale. The Pelican, while enjoying the response from others, is really a self-entertainer.

POLITICS

This is the hail-fellow-well-met who can spin a yarn or a speech at the drop of a hat. When this gift is combined with solid information (from his or her own knowledge or briefing by a staffer), this personable politician is hard to beat. Because s/he sees what is possible instead of dwelling on the vicissitudes in the political arena, s/he is reassuring to constituents that their government is in good hands.

FAIR FOWL

A fair Pelican has added a depth of thought and feeling to his or her repertoire. S/he uses his or her keen powers of observation of others to assist them when needed. S/he has added commitment to something beyond himself. With this greater capacity for joy comes a concomitant depth of sorrow. When the continuum is increased in one direction, it is balanced like a seesaw with emotions at the other end of human experience.

Because of his or her enormous charm, most other species do not feel threatened by a Pelican, so a gifted Pelican has protective camouflage. S/he is usually not the brunt of bullying by Snowy Egrets and Storks, or of manipulation by Cormorants. A fair Pelican can deliver wisdom with a wink and a grin; and those hearing can take the comment without offense. This ability to provide insights in a palatable way enables the fair Pelican to be an effective diffuser of differences and difficulties in a workplace or in a family.

A variation on this theme is the Pelican who plays Columbo, the fumbling detective, who is dumb like a fox. Pretending (with

his or her remarkable ability for drama) not to understand those who are combative or hiding something, s/he plays the country hick. There is a cunning here that keeps him or her in the Pelican range, as long as s/he is not malicious. One step over the line could define him or her as a Cormorant.

While s/he is the "down" that keeps everyone else fluffed up, periodically the fair Pelican needs others to hear his or her deeper concerns. It is not healthy to bob on the surface all of the time.

A Fair Pelican has learned to estimate the amount of time and energy that a project will take, and therefore, will be more realistic about his or her workload. Desirous of pleasing customers and yet aware that s/he must protect quality, s/he will negotiate a win-win arrangement when possible (or not accept the commission or job). Others would greedily agree and then waffle at delivery time.

The one thing I never learned to do in college was to create on schedule.
Agnes de Mille

In spite of this comment, Ms. de Mille did learn how to choreograph her energies in order to produce her remarkable repertoire of leading edge dances, working with some of the greatest musicians and dancers of the twentieth century.

FOUL FOWL

The Foul Pelican may be a shameless hedonist whose motto is "me first!" A Foul Pelican is particularly vulnerable to cumulative workplace fatigue; if s/he cannot leave the job, s/he may become sullen and unproductive. If things go wrong, s/he will feel unappreciated and become a complainer. S/he may become short-tempered, stubborn, easily angered, short with people, and given to tantrums. Intensely self-centered and selfish, s/he resembles a Foul Spoonbill.

Because of his or her own agenda, a Foul Pelican sometimes will try to protect or defend others and inappropriately insert himself or herself into a triangle. S/he needs to learn appropriate boundaries for all concerned.

The Foul Pelican can skirt the law if regulations are not

advantageous for him or her. Under normal circumstances, a Pelican is acquisitive. A dysfunctional Pelican will be at risk for bankruptcy, shop lifting, embezzlement, or failure to pay taxes. A slovenly personal appearance, yard, or apartment may be a clue to a Foul Pelican. Lacking self-discipline, a Foul Pelican can over-consume food, alcohol, and other substances.

STRATEGIES FOR INTERACTION WITH A SENIOR PELICAN

Most Pelicans are not interested in the power and requirement to supervise others that goes with seniority in a firm. They are more inclined to be entrepreneurs in fields that bring a sense of adventure. When one does encounter a senior Pelican, s/he is usually a fair fowl who has modified his or her behavior pattern with commitment to mission or to the company. S/he still will not cherish the bureaucratic mindset and processes, but will be on the lookout for adventuresome ways to make quantum leaps. Those who can contribute to such projects will be held in high regard by the senior Pelican.

A senior Pelican may believe that schedules are for other people (a belief also held by a Stork). However, the Pelican mindset is that organized meetings are not as productive as impromptu discussions that allow the flow to keep going. (The Stork is simply arrogant about letting others wait for him, even though that may be very demoralizing to his staff and very expensive in terms of time wasted waiting.) The Pelican will favor an open door, drop-in, management-by-walking-around style, and may not appreciate that other species find this disruptive and threatening.

A senior Pelican loves games. S/he plays to win, but is not obsessed with competition, as are some of the other species. Fun is the goal.

However, should this senior Pelican become stressed, s/he may be prone to outbursts, temper tantrums, and sarcasm (and seem like a Stork). Because a Pelican has typically not learned (nor valued) self-discipline, it is not in his quiver when s/he needs it. It is best to leave the Pelican alone at such times and let him or her

regroup. Although s/he may not apologize for the behavior, there is a tacit understanding that s/he lost control and face. One will have to read between the lines to understand what was going on with the Pelican because s/he will not want to tarnish his or her image by admitting personal struggles.

STRATEGIES FOR INTERACTION WITH A JUNIOR PELICAN

A junior Pelican can bring a great sense of creativity, laughter, and cohesion to a workplace. People enjoy congregating with the Pelican at lunch time because they know they will be fed more than the recommended daily dose of endorphins. The Pelican enjoys his or her audience. A wise supervisor will make sure that the Pelican has an opportunity built into the day to feed his or her pleasure factor. Otherwise, s/he will make the time, interrupting others' focused time.

There will be times when a supervisor of a Pelican just wants to take him or her and shake him. This desire to bring a stop and a focus to the Pelican is voiced by college professors as well as managers because they see the potential that they believe is being frittered away. A junior Pelican often can work well for a Stork because s/he has no ego in competition. A Pelican really has to be mishandled before s/he will push for his or her appropriate rewards. If the Pelican can be given a flexible schedule, s/he will be more productive than if forced to toe the nine-to-five time line. Because s/he does not tackle a project sequentially, a manager can wonder if the Pelican will be productive. The answer is usually yes, as long as the deadline is clear. If the Pelican can have a discrete portion of a project, that does not require on-going interface with others whose thought and work processes are different, s/he can be a very talented member of a team.

S/he is an innovator. S/he wearies of doing the same thing the same way each time. If the workplace can capitalize on the Pelican's dilettantism, variety will indeed be the spice. This is the chef who will change his or her recipe daily, just for the fun of it. Sometimes the result will be superb, and other times, not. S/he does not see consistency as a personal goal.

BOTTOM LINE

The Pelican can add to the bottom line of the workplace simply by his or her joyful approach to life, the fact that his antennae are always out scanning what is new, and her sheer energy. S/he has infectious enthusiasm that can enliven the whole environment.

Gather ye rosebuds while ye may,
Old Time is still a flying,
And this same flower that smiles today
Tomorrow will be dying.

Robert Herrick, *Hesperides,*
"To the Virgins to make much of Time" (1648)

Experts in what is lightly called the humor industry — organizational psychologists, authors, humor consultants — say our workplaces are in danger of becoming terminally serious. Lighten up on the job, they argue, and you can increase productivity, improve morale and boost the bottom line.

Julia Lawlor, "Employees encouraged to lighten up,"
USA Today (Sept. 23, 1991), B-1.

"Isn't my Grandpa great? He has more fun than rules."

"Dennis the Menace,"
by Ketcham (Feb. 17, 2001)

Probable Pelicans that I know:

____

____

____

____

Green Heron

Green Heron

Many people visit sites where a Green Heron is present, but they fail to see this remarkably camouflaged bird, even though he is the most widely distributed of the Herons. He is small (16 - 22 inches) and dark. With his chestnut neck and dark green, almost bluish/black feathers, he is hard to distinguish as he purposely avoids detection sitting in a fully-leafed shrub above the water. More chunky than the other Herons, he has a profile that looks like a Crow when he flies, but his deep wing beats will give him away.

When he is visible, often on a limb that leads from a shrub into the water, his short, greenish yellow legs and feet are the most colorful part of his body. In mating season, his legs and feet are orange. When he is surprised, he will stretch his neck (which is comparatively shorter than other herons' necks), jerk his tail, and a shaggy crest will appear. His voice is rarely heard.

His habitats are those of both fresh and salt water: lakes, ponds, marshes, swamps, and streams. He feeds from the muddy margin of a body of water, patiently motionless until he spots a small fish. Then, with great caution, a flick of his tail and the strike of his outstretched neck and bill, he has captured his unknowing prey. Small fish are his primary food, though crustaceans and small amphibians are also on his menu. On occasion, he will use one foot to rake through the mud to stir up salamanders or other small amphibians who might be hiding there. (The Snowy Egret uses one foot as well, but it is usually as he is standing in the water.)

Unlike all of the other species in this book, the Green Heron is capable of intentionally using "tools" to attract fish. For example, he may break off a piece of twig to use for bait. Or, he may use brightly colored leaves, flowers, feathers, or pieces of food left by humans, so that he is doing the fisherman's equivalent of "chumming." This activity is intentional, with the Green Heron retrieving bait when it floats away and carrying it to a new spot to try again. Also, unlike most of the birds in this book, he is usually a solitary nester.

GREEN HERON FOLKLORE

The Greek Goddess of wisdom was Sofia. Gnostics worshipped her. Agnostics follow suit today, challenging the knowability of the existence or nature of God and honoring the pursuit of knowledge for its own sake.

For the Romans, Athena was the goddess of wisdom and strategy. The daughter of Zeus, she emerged from the head of her father full grown, wearing armor and brandishing a spear. Her first sound was a battle cry. According to Jean Shinoda Bolen, M.D., author of *Goddesses in Everywoman,* she was known for her emotional distance, her craftiness, and lack of empathy. "Strategy, practicality, and tangible results are hallmarks of her particular wisdom. Athena values rational thinking and stands for the domination of will and intellect over instinct and nature." Not fond of change, unless she has initiated it herself, Athena keeps a tight rein on her emotions so that they do not complicate her intellectual pursuits.

Joseph Campbell, in *The Hero with a Thousand Faces,* tells the story of Daedulus, the creator of the labyrinth and also the provider of flax so that the hero Theseus could find his way back out of the labyrinth. "For centuries Daedulus has represented the type of the artist-scientist: that curiously disinterested, almost diabolic human phenomenon, beyond the normal bounds of social judgment, dedicated to the morals not of his time but of his art. He is the hero of the way of thought — singlehearted, courageous, and full of faith that the truth, as he finds it, shall make us free."

From the Native American tradition comes the Lakota Indian description of the fall season as a bear, an analytical thinker who can be indecisive. Of note is that fact that the bear hibernates for a major portion of the year.

OBSERVED GREEN HERON BEHAVIORS

The Green Heron thrives in environments like the mangrove at Tarpon Bay, Florida, in which there is an uncanny silence, broken only by the rare sound of a White Ibis searching for food. When searching for the Green Heron, one must be very still, looking for

eyes in the foliage of bushes near the water, or a mottled clump sitting on a branch that breaks the water's surface. He will see you long before you can spot him. He will be motionless for as long as you remain; only his eyes will move as you traverse the area. He is camouflaged, unflappable, and solitary. He is not in the company of others of his own species or of other species.

The only time his voice has been heard is before people have entered his domain, perhaps in delight in catching a fish or in frustration for failing to do so. He is not vocal like Seagulls, because he is solitary.

The Green Heron is a perching hunter, not a stalker. He will patiently watch a turtle swim by, hoping that his motions will stir up food for him. He is wary vs. aware. He is a master of masquerade. He knows how to turn his body to align with a root or branch and be practically impossible to be seen (by prey or bird watcher).

OBSERVED HUMAN BEHAVIORS

The Green Heron is often the scholar, researcher, or the intellect in the workplace. S/he favors a position in which s/he can work on a discrete program or project, preferably solo or with a few chosen associates who have respect for his or her mode of thinking. Generally, s/he does not want a leadership role that requires public interface, because that would take time away from valuable pursuits. (A classic example is the chronic avoidance of the department head position in universities where the job is primarily administrative and representational.) Because s/he finds most people less able, less serious, and an intrusion on his or her precious time, the Green Heron seeks seclusion as much as possible.

If the workplace is sufficiently staffed to permit this strategic thinker the opportunity to proceed without micro-management, the results may be remarkable. While others tend to be involved in the day-to-day challenges, the Green Heron tends to seek quantum leaps in operations and knowledge of all sorts. S/he is an inventor of concepts. In order to have the luxury of letting his or her mind pursue these options, s/he requires a calm, quiet setting that is

rarely invaded by requirements for immediate reports, public discussion of a current project, or attention to mundane details that keep an office or home functioning well. Grant applications, necessary for funding one's research, are particularly heinous.

A Green Heron dislikes a one-size-fits-all approach in employee rules and regulations and in corporate decisions. S/he dislikes large, indiscriminate categories because they do not sufficiently differentiate. For example, a Green Heron would dislike the metaphors in this book, because the clumps are too large to describe the variations within each behavior pattern. The Green Heron dislikes metaphors, but often uses similes: something "is like..." This seems like a fine line, but Green Herons are fond of fine lines.

A Green Heron sees grays in the black/white continuum, especially at the darker end. S/he operates on a win-lose basis, but the playing field is specifically an intellectual field. A Green Heron will contribute significantly to corporate or military planning if given the opportunity to listen and ponder. When s/he is ready to speak, a remarkable assessment will be rendered quietly and in relatively few words. In settings where the most aggressive or the noisiest gets the most attention, the Green Heron's contribution might not be recognized. S/he does not see the need to take a popular position. The wise leader will provide an environment in which this apolitical creature can be valued and nurtured. The Green Heron does not see dissent as disloyalty.

The office of a Green Heron will often seem chaotic to others, but s/he knows exactly where the information is that s/he needs. With the advent of technology, the Green Heron may spend what seems like an inordinate amount of time on the computer. A lover of data and organized management of that data, s/he finds comfort and power in the accumulation, manipulation, and analysis of data that would overwhelm others.

Because s/he is so intensely focused on cognitive operations, interruptions by others can be met with an irritable, almost surly response. Whether those interruptions be for corporate tasks that all employees are expected to participate in, or personal queries,

the result is a breaking into a private world. A Green Heron prefers not to be reachable by phone, pager, or a beeping computer that announces, "You have mail." S/he wants to choose when and where such intrusions are permitted. Obviously, an open door would be an anathema, as is management by one's superior's walking around.

Arenas that require stealth and secrecy are comfort zones for a Green Heron. Assurance that one's peers have appropriate credentials in order to participate in a project helps a Green Heron be somewhat less vulnerable, but s/he will not be sticking out his or her neck (or ideas) until s/he has a complete solution, if s/he can avoid it.

While Athena was considered ambitious, the Green Heron's ambition is not for a public profile. For a Green Heron, authority is derived from knowledge, not from position, so s/he can find status leaders incompetent and not worthy of being followed. Allegiance is to an abstract, such as knowledge or truth; loyalty is not personal. His or her ambition has more to do with cognitive or scientific breakthroughs. While a Green Heron is often overlooked at corporate award time (with recognition going more often to those who market themselves and their accomplishments well), his or her award is really an internal pride in an elegant solution to a problem or the identification of a problem no one else has even noticed. And the greatest prize would be being allowed to keep working at his or her own pace and in her chosen intellectual domain.

The Green Heron tends to see everything in complex terms and is irritated by those who do not (either because they are shallow or because they are visionary and see simplicity beyond the complex). Whenever one asks for a short explanation from a Green Heron, be prepared for more than was needed.

An introvert, the Green Heron intensely dislikes staff meetings, conferences (at which little of any significance is really shared), and social gatherings of any kind. Sports and exercise tend to be those that can be done alone (to permit mulling time) or with a few chosen others (and there will be little or no conversation). Team

sports and corporate teamwork are not favored venues. Nor is the Green Heron drawn to service organizations where friendship and networking are at least as much a draw as community projects. Organizations which accentuate individual achievement, such as the Boy Scouts' Eagle Scout award or the Girl Scouts' equivalent, will capture the interest of a Green Heron for awhile, but s/he is unlikely to be a leader as an adult.

Interestingly enough, the Green Heron may often appear as a daredevil in his or her youth; but, on close examination, the risks have been carefully calculated based on skills and physics. These seemingly hysterical antics are not in the same ball park as those of the Stork or the Pelican. The Green Heron may rescue a downed pilot, but he will do so having analyzed the situation vs. simply going with a flow of adrenaline. His response to the gratitude and admiration of others will be embarrassment for his act having been made public.

A logical thinker, s/he distrusts presenters and proposals that do not have a sequential flow. Emotions and anecdotes have no place in formal presentations. Neither do repetitions, self-serving claims of ownership or discovery, or inflated claims.

When determining courses of inquiry or action, the Green Heron studies all options and feels no compunction for haste. Because of his or her desire for the best solution, s/he may be overcome by events or by others taking the ball and running with it. What looks like procrastination to others, because the Green Heron does not feel compelled to communicate his or her decision-making process, may indeed produce an innovative initiative. Procrastination may mask persistence and perseverance. The luxury to take time with each decision, however, is not the norm in many workplaces.

S/he is critical in his or her refusal to be a "yes man." In many settings, the Green Heron plays the role of devil's advocate or the critical commentator. S/he is thought provoking and, sometimes, just plain provoking. Because of his or her intellect, few enjoy being queried by a Green Heron. It feels cold, condescending and, without softening interpersonal nuances, abrasive. While many

Americans value the suave, considerate, flattering and charismatic charms of other birds, the Green Heron values mental and social sandpaper. A Green Heron can do the equivalent of the fowl's chumming — tossing out bait and waiting for the unwary or self-focused to bite.

A Green Heron makes careful, wary eye contact with others, often with an almost twitching motion. S/he really prefers to observe others while not being observed himself. Generally, the Green Heron will not be confrontational. S/he will avoid conflict whenever possible. S/he will simply sequester himself or herself, becoming uncommunicative and hoping the tensions will dissipate. Only when startled or trapped, will s/he be combative.

S/he does not like being touched. While some of the other birds thrive upon comrade punching, bear hugs, and familial hugs and kisses, the Green Heron finds these an effrontery.

Because the Green Heron dislikes collaborative endeavors (because they honor the input of all members, regardless of their perceived value), s/he may sabotage collaborative initiatives. A collegial setting, in which workers operate as peers and work on their own projects, is usually a more effective arrangement for a Green Heron than a hierarchy or a collection of teams.

VISIBLE SIGNS

The Green Heron's office is an inner sanctum. S/he really does not enjoy company in this private space. It appears chaotic, partially due to the sheer volume of papers, books, or equipment tucked into all available nooks and crannies. The floor becomes a horizontal file when s/he runs out of space elsewhere, so navigating the room is difficult, if not hazardous. S/he saves everything and can usually find it when s/he needs it. The Green Heron's office often needs a breath of fresh air and a thorough cleaning (which s/he avoids because someone else might move his or her stuff and s/he does not want to waste time doing it herself). Not focused on appearances, s/he simply is not aware and does not care how tired and worn the furniture is and how aesthetically displeasing his or her environment has become.

The Green Heron is not a stylish dresser. Again, camouflage or at least a low profile is desired. Sometimes, like the absent-minded professor, s/he cares so little about appearance that clothing is mismatched or soiled. Clothing is utilitarian. Green Heron professors or teachers will wear the same suit all week, with a shirt or blouse change as the only apparent difference. If s/he can wear a uniform (like a laboratory jacket or a military uniform), s/he likes not having to waste time thinking about what to wear and likes knowing who's who by the uniforms they wear.

HUMOR

The Green Heron enjoys puns, intellectual ironies, and teasing those with whom s/he feels comfortable. Just as the bird has orange legs during courtship, the human version can exude a brilliantly endearing humor when s/he feels safe in shedding his or her distanced mask.

S/he may use sardonic wit like a skewer in situations in which s/he sees puffery, profanity, intellectual pilfering, or personal pandering. S/he does not suffer fools gladly.

Political satire and commentary delight a Green Heron, especially of the "Doonesbury" and "Dilbert" genre.

POLITICS

A Green Heron is a rare figure in politics, because s/he dislikes the public profile that comes with the territory and dislikes the simplified, polarized arguments. However, when s/he does run for office, it is usually on the basis of deep commitment to specific causes and with a significant intellectual base. A Green Heron tends to be a pessimist vs. an optimist, which is not a popular stance with many Americans. S/he sees the complexity and ambiguity in our national and international affairs and often lacks the ability to translate his or her platform in terms that the voting public understands. Bill Bradley is a recent example, as is Al Gore. Adlai Stevenson fit this profile as well.

S/he is interested in the grand scheme of things; s/he is often a history buff in a specific field. This fits nicely with his or her delight in time alone, reading, researching, and pondering.

A Green Heron voter tends to be conservative politically and basically is not sympathetic to the concept of "Give me your tired, your poor, your huddled masses yearning to breathe free." The homeless and tempest-tossed are not where the Green Heron wants to expend his or her interest and concern.

FAIR GREEN HERON

A fair Green Heron has added a personal warmth and consideration to his or her intellectual bias. Where a Green Heron would rarely be selected as a natural leader (though s/he may gain status leadership due to his or her expertise), the fair Green Heron understands the value of gentle humor, approachability, and praise for others' work. Although these characteristics drain energy, and s/he must have private time to regroup, s/he believes that the effort is critical to the accom-plishment of the mission. S/he has learned how to balance his or her time alone and time with others, so that those around him or her can enjoy full focus, albeit for short spurts of time. S/he still needs solitude to replenish and nurture the cognitive life.

The fair Green Heron often finds music a beneficial avocation. Drawn to complex music, often styles that feature variations on a theme (such as chamber music and jazz), the Green Heron enjoys the technical more than the expressive aspects. Woodworking is a solitary occupation that a Green Heron finds enjoyable. Because s/he is often capable of doing two things at once, one active and one cognitive, this is an attractive pastime (while many fix-it chores remain untouched).

The fair male Green Heron can be a charmer when he wants to be, albeit somewhat awkwardly. He uses his diplomacy or gallantry from a position of power.

FOUL GREEN HERON

Deprived of positive human interactions with appreciative others, the Green Heron can become a bitter, surly cynic. Unaware and/or uncaring about his or her impact on others, s/he uses the heavy duty/coarse grade of sandpaper, when a fine grade would have sufficed. Others avoid him or her whenever possible, which ostensibly suits him just fine. But, at some point, the sense of isolation and victimization can overwhelm the Green Heron.

A little privacy is one thing; isolation is another. Not used to providing the basic niceties for himself or herself, the Green Heron can find that no one is making coffee, offering to pick up lunch, typing his or her reports, or connecting in the myriad ways that make life easier and more pleasant. This happens in families as well as workplaces.

A foul Green Heron becomes a "No-man," exceedingly pessimistic in his or her outlook and discouraged from productive work. In workplaces, the foul Green Heron may be handed a pink slip if corporate leaders find his or her abrasiveness unpalatable and unproductive. Leaders will be alerted to the malfunctioning Green Heron by the departure or requests-for-transfer by co-workers (or by the volume and nature of equal employment opportunity complaints). They will also be alerted by stoppages in work flow when materials reach the Green Heron's desk. Simple operational tracking can pinpoint this problem.

STRATEGIES FOR INTERACTION WITH A SENIOR GREEN HERON

When working for a Green Heron, it is important to keep your focus on the work at hand, not on personal interface. S/he often will resist attempts at any personal connections. Some call the female version an "Ice Queen." Make an appointment to discuss a specific agenda. Make your materials logical, concise, and accurate. Understand that questions about your work are usually not personal; they are an attempt to validate your processes or information. Questions may sound haughty, when in truth they are an attempt to capture your insights, which you may have gained through processes foreign to his or her way of thinking. It

can be uncomfortable and unnerving, but if you know that your work holds water, try not to over-react to his or her interest and scrutiny.

Over time, your Green Heron may find you trustworthy and may depend upon you to be his or her public face, thus ensuring warmer relations with other departments without infringing on his or her concentrated efforts. S/he may also depend upon you to explain his or her work to those who need the information in order to complete their assignments. You may be able to put the concepts in context for those who must use them and have the patience to be the point of contact as they adapt them in their settings. This frees your Green Heron boss to explore new domains.

If your senior Green Heron baits or skewers you, it is important to let him or her know (privately and as soon as possible) that that behavior makes you less productive. If you do not address it promptly, s/he may not have any memory of what you are talking about, because s/he really does not store that kind of information. The fact that your feelings might be hurt generally does not compute either. What matters is that you cannot perform as expected.

STRATEGIES FOR INTERACTION WITH A JUNIOR GREEN HERON

A junior Green Heron may be given some latitude in terms of blocks of time to pursue discrete projects. However, in order to preclude the development of a foul Green Heron, it is important to require his or her appropriate interface with co-workers and with you. S/he may be a "genius," or at least aspire to be one, but s/he needs to hear what projects others are working on, what problems they are encountering, and what impact the larger economy/ environment is having on your operation so that s/he can be attuned and using his or her talents most productively for your mission.

Your junior Green Heron also needs deadlines for completion of some portions of key projects. While s/he will resist this, it is

important that s/he does not get stuck perfecting a minor phase while the global focus requires moving on with the "best possible" for now. S/he must experience the consequences for deadlines missed.

There is a tendency for a Green Heron with expertise in a given field to become egotistical and arrogant. S/he needs to learn to appreciate what others bring to the table. S/he needs to become acculturated to your corporate environment and not be allowed to operate independently. A Green Heron wants the dichotomy of power and being left alone without direct responsibility. S/he really does not want to be out on the playing field with Storks, Blue Herons, or even other Green Herons. S/he wants to be able to be free from the contamination of power plays and compromises, but wants to be able to snipe from the sidelines with impunity. S/he believes she is above the battle and is "cognitively purer-than-thou."

When you spot incidents of baiting, sardonic skewering, or surly behavior, it is important to tackle them immediately, preferably in private, and make clear that they are unacceptable in your workplace. Allowed to get away with these nasty performances, the junior Green Heron could be well on his or her way to a foul state — a loss for your workplace on many levels.

As a supervisor, you may need to protect a Green Heron from others making hasty requests or attempting to "pick his brain." The Green Heron prefers time to mull and give a complete answer later. A supervisor may need to prioritize the requests coming in for the Green Heron's sake. S/he likes check-off lists and may even draw boxes around separate concepts in order to keep the compartments clear. The Green Heron is particularly fond of compartmentalization. His or her life is in his mind, inaccessible to most others.

BOTTOM LINE

Steer by the stars, not by the lights of passing ships.

General Omar Bradley

Choose something like a star...
We grant your loftiness the right
To some obscurity of cloud...

So when at times the mob is sway'd to carry praise or blame too far,
We may choose something like a star to stay our minds on and be staid.

Choose Something Like a Star, Robert Frost

Ask the impertinent question, because only by so doing will you get the pertinent answer.

Gary Trudeau,
Commencement Address, Smith College, 1987

Probable Green Herons that I know:

____

____

____

____

Double-Crested Cormorant

Double-Crested Cormorant

The Double-Crested Cormorant is the most widespread Cormorant in Eastern North America and seems to be increasing in numbers in most of his range. He is a big bird, ranging from 30 - 36 inches tall, with a wing spread ranging from three to five feet. He is found both on the coast and inland, inhabiting both salt and fresh water bodies of water. He is black with an orange throat pouch and a hooked bill. His typical pose is spread-winged on a buoy, piling, rock, or pier, as he dries out after a period of prolonged underwater fishing.

His crest is rarely visible, probably because it is wet. His feathers are not quite water-proof; so a major vulnerability is his need to find a tree, an island, or, in a real emergency, a beach on which to come ashore and dry out. Because of his short tail feathers, he can sit on sand or low mangrove roots; but, his relative, the Anhinga, must find a tree because of his long tail feathers.

He is a ravenous eater, the scourge of inland pond fishermen. He slants his bill upward while swimming rapidly through the water. He can expand his mouth and esophagus in order to consume large fish. His webbed feet give him great speed and maneuverability underwater. He has been used by Japanese fishermen to help catch fish. This requires a chain around his neck to preclude his consumption of his catch; the fisherman then determines what share he should have. This is not a voluntary or domesticated relationship.

His salt water diet is primarily fish, eels, and crustaceans; in fresh water, he will also feed on reptiles, insects, and amphibians. He pursues most of his prey underwater, but must surface to consume his catch. He will also steal fish from another Cormorant's mouth. He is gregarious, often seen with others of his own species as well as those of other species, both for feeding and resting.

CORMORANT FOLKLORE

Maryjo Koch, in *Bird, Egg, Feather, Nest*, gives this history: "The Cormorant is shrouded not only in dark, dusky feathers, but in human bias as well. Deemed 'unclean' in Deuteronomy and synonymous with 'gluttony' for Chaucer, the Cormorant has also been defamed by Milton and Shakespeare with their allusions to Satan and the 'Devourer of life.' ... The word Cormorant means 'Crow of the sea,' but this big black water bird is closely related to the Pelican family... and dries its feathers in a spread-winged Dracula pose." When he is shown looking left, this ties with the symbolism of the sinister (literally, sinistre, meaning left).

History teaches us many lessons about Cormorants, including the Inquisition and witch hunts. Cassius is a classic conspirator in Shakespeare's *Julius Caesar,* and Alexander Hamilton undermined his commander-in-Chief and President, George Washington. In David McCullough's book, *John Adams,* it is clear that George Washington surrounded himself with Storks, who are not really good judges of character. John Adams let Hamilton go unleashed far too long. The dilemma of identifying someone clearly as a Cormorant is the basis of much fine literature. Nathaniel Hawthorne's *The Scarlet Letter* challenges readers to determine if young Pearl, the fruit of an immoral pairing, according to Puritan ethics, is capable at her early age of being "evil." In The *Crucible,* written by Arthur Miller during the McCarthy hearings era, we see an historic situation being used as a symbol for modern excesses.

Satan is a fallen angel in the Bible. As Milton discovered in *Paradise Lost,* the evil one is innately more interesting than the saint. Literature is full of antagonists that intrigue the reader more than the protagonists. According to Burrows, Lapides, and Shawcross in *Myths & Motifs in Literature,* they are often rather "sophisticated gentlemen who can talk a person into leaving the path of righteousness." *The Devil and Daniel Webster* is a classic.

The temptress who ensnares would-be heroes, detains and detours them from their noble quest, or destroys them is another Cormorant motif in literature. *The Iliad* and *The Odyssey* provide lessons on this theme, as does the serpent in the Garden of Eden.

Cain is the earliest human Cormorant in the Bible. He asks God if he is expected to be his brother's keeper. Underestimating God's omniscience, he attempts this ploy when he has already killed his brother Abel.

King Herod is the Cormorant in the Biblical story of the birth of Jesus. Looking to gain information from the Three Wise Men, he cunningly encouraged them to visit on their return to tell of what they had seen. They wisely chose another route. However, Herod was enraged and ordered the slaughter of the innocents, all children who might be in the age range of the Holy Child. Jesus was betrayed by Judas Iscariot, another Cormorant.

Joseph Campbell, in *The Hero with a Thousand Faces,* relates tales of the trickster, usually a Raven in Eskimo mythology, who is sometimes admired for his cunning; but, he causes harm (unintentionally) by his curiosity or greed. Then, he must use his wits to prevail. He sounds much like the Cormorant because the Eskimos of the Bering Strait describe him as sitting, drying his clothes on a beach. In another tale from the coast of Africa, the greatest joy of Edshu, the trickster god, was spreading strife. Graeme Gibson, in his book *The Bedside Book of Birds – An Avian Miscellany,* notes that Ravens have "figured out a way of bringing predator and prey together."

The symbolism of two faces is particularly apt for human Cormorants, who use their playful side to camouflage their dark side. Often, those who are victims of Cormorants are perceived as the "bad guys" who persecute the poor Cormorant, so the Cormorant has a double win yet again.

The Cormorant can be a Jekyll and Hyde, being very selective in her mal-behavior pattern. There may be a <u>public persona</u> that fits another bird and a <u>private persona</u> that is pure Cormorant. S/he often sees herself or himself as an actor or actress.

OBSERVED CORMORANT BEHAVIORS

The Cormorant is found even in very small bodies of water frequented by numerous other birds, such as the Wood Ibis (Stork) and White Ibis; he slithers under water taking up far more of the

pond than his vertical comrades. He pokes his head out for air and to check the surface quickly, and then re-submerges. He grabs his prey in his eagle-like bill. Technically, he is not a raptor like the Eagle, Hawk, Vulture, Owl, or Osprey. His bill has the same shape, and he has big nails on his feet, but he does not tear apart his prey. He swallows it whole. When he swims on the surface, with only his head and neck out of water, he resembles a swimming snake.

A Cormorant thrives in closed ponds. When the Cormorant feeds in the ocean, he must emerge to dry out. If he does so on the beach, he is vulnerable during the period that he cannot be airborne or underwater. His takeoff for flight is clumsy under the best of circumstances, and is terribly cumbersome when he is wet and threatened. His sheer greed keeps him fishing longer than is wise, and then he must dry his wings.

A Cormorant is so tenacious in pursuing his prey that he sometimes forgets where he is. One particularly amusing episode featured a Cormorant chasing his prey right up onto the shore of a small stream, right into the feet of human observers. He looked a little chagrined, if a bird is capable of that emotion! He lost his prey and exposed himself to potential danger.

Often you cannot see a Cormorant under water, but you can see small fish jumping to avoid him. When he is visible in shallow waters, he looks like a stealth plane, with his wings outspread. In deeper waters, he propels himself like a missile with his webbed feet.

One remarkable experience with a Cormorant came as a Seagull pursued him relentlessly. Having seen that he had caught a fish on our pond, the Seagull kept after him every time he surfaced and tried to swallow his catch. The Seagull forced the Cormorant to take a very short breath and submerge again. Each time he came up for air, the Seagull pounced. The Cormorant finally gave up the fish to the Seagull.

A Great Blue Heron will also finally take on a Cormorant that is harassing him. The strike by the Great Blue Heron's bill is lethal.

A particularly unusual symbiosis is the paired flocks of Cormorants and White Pelicans who fish together. The Pelicans steam along in a clump or a line, acting like a feathered net. The Cormorants fish underwater in and around the Pelicans, being delighted with the prey trapped by the white armada.

The Cormorant is not made to soar; he flaps his wings in a pattern different from the others who have more soar and glide capacity. His real strength is in his underwater swimming prowess.

The Cormorant coughs up pellets of undigestible bones of his prey. So, there is visible evidence of his activity. It would be helpful if there were a human parallel! "Cough it up!" would take on new meaning.

OBSERVED HUMAN BEHAVIORS

The Cormorant, in our context of behavior patterns, is EVERY ONE OF THE PATTERNS GONE AWRY (with the exception of the Scurry Birds and the Skimmer). Most typologies do not provide for a separate category. However, my observation indicates a need for this separate entity, which retains the feathers of the original fowl. The key designator is the malicious desire to harm others. All of the birds and behavior patterns irritate others in some way; but, that is just part of their profile. When the hurtful behavior becomes intentional, then the individual is a Cormorant.

Determining that intention is key. It is also very difficult; particularly in families and in workplaces, as well as in all of the other environments in which we encounter people. We must assess the motivation and the repetitive nature of the actions, because one negative slip does not equate to the cosmic fall from grace. It may take years to feel sure that what you have experienced has not been random or simply the obnoxious outfall of foul fowls. On the other hand, you may have a flash of insight that clearly defines the individual as a Cormorant. The Cormorant stands for all foul fowls who have stepped over the line of normal behavior and are purposely seeking to harm others.

The question arises frequently, can people change their feathers? Clearly, in the domain of ill intent, yes. In the positive

domain, can one become another fowl? Probably not, but one can become a fair fowl within the behavior pattern, and can emulate behaviors that s/he admires in other patterns. This takes a great deal of effort and, therefore, demonstrates a great desire to modify one's usual pattern. Martin Seligman says, in *Learned Optimism,* that while optimism has a genetic component, it may be learned at any age (though it is harder as one gets older and has a lifetime of behavior habits to change).

The Cormorant uses polarized thinking, such as either-or, win-lose, and black-white (mostly black). There is no room for multiple interpretations or layered thinking. A very concrete thinker, s/he is confused by metaphors. S/he sees the world as finite, and s/he is not getting her share. This attitude shows up in widespread bigotry. S/he operates on a win - lose or a lose - lose perspective. S/he is willing to lose in some ways if by so doing s/he damages others in the process.

The defining characteristic of a Cormorant is his or her quest to find a mistake or wrong-doing by his quarry and then s/he can play "gotcha." S/he is a hunter....but the motivation is to gain power over those who have power (or something else) that s/he envies. If one befriends a Cormorant (or not), there will be a price to pay for this relationship. In the workplace, s/he is like a cancerous tumor that festers, spreading ill will where there has been delight and camaraderie. The damage that one Cormorant can do to your favorite fishing hole is unbelievable, between his nasty habit of fouling the very waters in which s/he fishes and his consumption of all available food (translate "resources").

Because the Cormorant is not capable of introspection, s/he feels no remorse for his or her actions. This, added to the intentional hurt inflicted, makes a Cormorant very dangerous. S/he is committed to find the flaws in others, because if s/he fails to do so, s/he might have to look at herself — which s/he cannot bear, unlike other birds who thrive on self-awareness. The Cormorant's attacks really come from his or her own baggage, her load that is too loathsome to examine. S/he needs to dump it somewhere.

The Cormorant darkens the environment s/he inhabits, physically and psychologically. S/he literally enjoys a darker office lighting than usual and would prefer to be allowed to work (unsupervised) after hours or before hours. S/he brings an emotional gloom much of the time, except when in his or her methodically playful mode, which is diversionary in nature.

As in abusive relationships of the domestic variety, there is a specific cycle of tension-building, violence, "remorse" (which is usually of a shallow nature), and a honeymoon period before it all starts over again. The clincher is someone else saying, "S/he's doing it again." This is a shorthand that may be the first confirmation that you, the victim, are not to blame.

The Cormorant likes secrets. S/he will purposely tell you a secret and then try to force you "not to tell." This is a nasty form of control and is designed to make you a co-conspirator. Be prepared to say, "I don't want to hear secrets and I will not keep them."

Game playing of all sorts is a favorite pastime for a Cormorant. S/he makes up the rules and s/he cheats. S/he relishes the excitement of uncalculated risk, while other birds prefer to avoid it.

Volatility and changeability are characteristic of a Cormorant. S/he may be headed in one direction and then is off on a different tangent, with accompanying extreme mood changes. Typically, s/he has high energy, both mentally and physically. S/he connives 'til s/he drops!

A Cormorant knows all of the rules and uses them when they suit his or her purposes to avoid being caught or to catch others doing wrong. Otherwise, s/he breaks the rules. For example, the Cormorant leaves work when s/he chooses, exiting by a back door or pretending to be making a coffee run. Because most others are focused on their own work, they are not keeping tabs on the Cormorant, who takes advantage of an environment of trust. One would think that a Cormorant might be especially effective in a spy/intelligence role because of his or her delight in subterfuge. However, s/he is unreliable.

It is particularly sad to learn that a Cormorant parent will tease, provoke, or otherwise push his or her children until they misbehave; then the Cormorant may punish without compunction. So, when the Cormorant is faced with adult children who have learned to handle their dysfunctional parent, s/he may indeed become even more dysfunctional in order to elicit illicit response from her baffled adult children. The Cormorant-in-extremis baits even those who have learned to manage earlier strategems. Absence and distance are the solution here.

Technically, s/he is not a raptor, but s/he is rapacious, which means excessively grasping or covetous; given to seizing or extorting what is coveted; ravenous; voracious. S/he is a predator. Many of us have been raised to be respectful of others and to believe that folks are not intentionally hurtful. We need a new paradigm for the Cormorants in our lives, whether they are a parent, sibling, spouse, teacher, or someone in our workplace. A Cormorant chooses as victims those least likely to confront him or her head on. Some Cormorants have been abused as children or as adults. While their victimization may help us understand, it does not excuse their behavior, which may include harassment (sexual and other). Greed is also a motivator in corporate settings.

There is a distinct difference between human Storks and Cormorants. A Stork is openly, unabashedly aggressive, loud, and usually surrounded by others like him all of the time, but is not malicious. Whereas, a Cormorant is subversive and malicious. S/he usually feeds solo, though there may be others around when the food supply is plentiful or when they stand together for protection while they dry their wings.

A Cormorant will use wiles and cunning in her pursuit of her prey. Variability of attack is one major characteristic of the Cormorant. S/he may be playful and coy one moment, and then be critical, perfectionistic, and caustic the next. If very sick or vulnerable, s/he can regress to a "poor me" dependency; but you will know when s/he is feeling better!

VIOLENCE IN THE WORKPLACE AND SCHOOL SETTINGS

Just as school personnel are coached on the warning signs of students who may become violent in a school setting, workplace personnel must heed the often subliminal alerts that they sense. Once a Cormorant is identified, many wonder how they could have missed the indicators that they recognize in hindsight. The truth is that a Cormorant does break the surface periodically, but it is not enough on any one person's radar to be spotted easily. The breakthrough usually occurs when the Cormorant has gotten overly confident and careless.

Cormorants are becoming more brazen in workplaces today, under the guise that current total quality leadership philosophies and processes encourage employees' constructive critical thinking. However, when the behavior becomes blatantly critical, arrogant, aggressive, and accusatory of seniors and others, then the Cormorant is undermining the intent of TQL, which emphasizes responsibility and respectful interface. Sometimes several Cormorants join forces, recognizing their strength in numbers, but also hoping to counter fish-snatching by their fellow Cormorants. So, it is an uneasy alliance.

In bureaucracies, government and corporate, where complaints are highly regulated, the Cormorant has learned how to use the EEO process to his or her advantage. When the Cormorant fears a lowered performance rating, s/he quickly muddies the water with an EEO complaint. Even if it is blatantly frivolous, the complaint serves to focus unwanted and unwarranted attention on the supervisor or other target. Because the bureaucracy wants to remove the blemish of the complaint, regardless of its validity, pressure is applied to meet the wishes of the Cormorant, whose laundry list always includes the reversal of the existing or impending low performance rating.

This is not to imply that there are not bona fide EEO complaints. But, like all processes, they are tools in the hands of various users. Hot lines can be manipulated by unscrupulous callers as well, with resultant witch hunts.

The track record for those who use the EEO complaint process frivolously is that as one case draws to a close, s/he initiates another. It is addictive behavior that gives the initiator a vindictive high (and lots of attention).

Cormorants are also more brazen in school settings today. The national concern with bullying has uncovered not only the physically and sexually abusive bully, but the vicious acts by girls: "They try to find your weaknesses...They won't hesitate to openly mock your physical flaws... If that doesn't work, they'll try to destroy your reputation... I'd rather be bullied by a guy any day than by a girl." These are comments by high school students related by Patrick Welsh, a teacher in a Northern Virginia high school, in an article in *USA Today*. Rumor, innuendo, character assassination, ridicule, and the silent treatment are their weapons. Rumor is a particular favorite because it is submerged, hard to trace, and usually gets exaggerated (a la the old telephone game).

CORMORANT PREY

Who is the prey of a Cormorant? Anyone who has something or accomplishes something that the Cormorant covets. A Cormorant preys on those who make him or her feel less worthy. Not being willing to compete above board, s/he operates below the surface, sabotaging in every way imaginable. There often has been no intentional, direct hit by the Cormorant's victim. But, the Cormorant has perceived the other's success, attractiveness, or openness as a threat.

When attacks come absolutely out of the blue, with a lot of emotional content, the victim should examine the flock. This attack is not about what the victim has done; it is all about the Cormorant. The vitriolic nature of the behavior is the give-away. No other fowl operates this way. Often, the target is someone who is a decent individual who believes that most people operate from a position of good will. A Seagull or a Great Blue Heron are favorite targets, because of their positions and behavior patterns. A Cormorant will rarely tangle with a Stork or a Snowy Egret.

However, the Cormorant underestimates the Seagull and the Great Blue Heron. Observations in the natural and human world

bear this out. Unfortunately, s/he can cause a great deal of havoc before they finally are sufficiently aware of activity and motivation.

Full of guile, the Cormorant is always focused on winning something away from another. So, while most others are oblivious to his or her conniving, indeed focused on their mission, the Cormorant has a single purpose to be achieved whenever opportunity presents itself or when s/he makes the opportunity.

The Cormorant is truly reactive; s/he must have a target. If for some reason, the target leaves the scene, s/he must find another. A truly dysfunctional Cormorant will be devastated physically and/or emotionally by the loss of the target that has become the sole focus of his or her existence.

Those who are not directly impacted by the Cormorant may simply see him or her as flaky, playful, or colorful. His or her antics are amusing. S/he is seen as audacious and "a piece of work." But, to the prey, the Cormorant is sometimes reckless in pursuit, just like the Cormorant who ended up on shore when chasing minnows. S/he is so focused on obtaining his or her desire (whether that is a relationship, job security, or a scheme of some sort), that s/he will take action without sufficient camouflage, will attack the reputation of others, and will give away information without realizing it.

Because the Cormorant sees himself or herself as the center of every thought or action of others, s/he is an extremely sensitive fowl constantly scanning the horizon to see how s/he will be affected. This hypersensitivity leaves him or her emotionally raw and drained; and s/he focuses on what s/he perceives as the source of hurt with an intensity that can become a raison d'etre. S/he does not want forgiveness; it infuriates him or her. True self-knowledge is lacking, but observance of others with a distorted refraction is part of his or her hyper-vigilance.

CORMORANT STRATEGIES

Abuse by a Cormorant can be verbal. It is all about control. The actual words may appear innocent enough, but the tone or the

accompanying body language are the give aways. For example, when a Cormorant responds to news of someone else's achievement, s/he says, "S/he did? vs. the usual emphasis: "S/he did?" A seemingly slight difference, but it is a slight for sure. If one has a parent, teacher, or boss who uses such a stratagem consistently, but transparently, one could internalize the subliminal message.

Another verbal strategy is to prompt a compliment for someone else, hoping to gain some reflected glory. When seeing someone else in the limelight becomes too painful, the Cormorant will make derogatory comments just loud enough so that s/he is sure that the complimented target can hear it. Then s/he will turn furtively to see if the hoped-for effect is occurring. The strategy for the injured in this scenario is to ignore the Cormorant in public. Later, decide if it is worth taking him or her to task privately as s/he will claim that the target heard it wrong and will relish the impact of getting "the dirt" out there twice, doubling his or her pleasure.

CORMORANT SUBTERFUGE AND SABOTAGE STRATEGIES BY ORIGINAL BEHAVIOR PATTERNS

The **Egret/Cormorant** supervisor quizzes a subordinate until s/he entraps. S/he will ask the same question 18 times until the subordinate wearies and wiggles on the set response in an attempt to move the discussion along. Now the subordinate is dinner. Because both the Egret and the Cormorant focus on flaws in others (certainly not on their own), one has to assess carefully whether one is encountering a foul Egret or a Cormorant.

Another version of this requirement to "prove it," coming from a **Green Heron/Cormorant** or **Stork/Cormoran**t supervisor, is not due to intellectual analysis, but comes from a personal experience with being burned by a lack of proof. The effect is that the subordinate is stupid, guilty, or under-handed until s/he, the Cormorant, determines otherwise. This Cormorant is a very defensive individual and it is very hard for co-workers to build up enough credibility/loyalty to compensate for his or her hurts. The

well-known ratio of ten compliments or successes to balance one negative does not begin to capture the distrust here.

The Cormorant is not capable of true compromise (because s/he operates on a win-lose or lose-lose philosophy). S/he will see a compromise as a loss and will subvert any agreement s/he is forced to make.

The **Egret/Cormorant** gives charitable gifts with onerous strings attached, specifying uses of the funds, requiring reports on their use, and generally entwining himself or herself in the business of those to whom s/he has been "generous."

The **Stork/Cormorant** engages other Cormorants to serve as moles in an operation or corporation. The ability to manipulate venomous activities gives him or her a great sense of power. Cormorants are especially vulnerable to manipulation by other Cormorants and other species who do not have the courage to do their own dirty work. When they can no longer reach their target, they are likely either to shift to a new target with renewed venom or turn their venom inward and self-destruct.

The **Stork/Cormoran**t is spoiling for a fight. Keenly aware of every slight that s/he perceives s/he has received, s/he has a chip on his or her shoulder a mile high. "You never listen to what I tell you!" is a leitmotif. In spite of his or her vocal bravado, s/he is uncertain of herself in many ways, so s/he bluffs physically and otherwise. S/he is very confrontational physically, verbally, and socially, honoring few of the niceties or even regulatory requirements. Subtlety is not part of his or her cognitive vocabulary. Because s/he may wear a weapon and a uniform, s/he is intimidating.

A **Stork/Cormorant** will brazenly enter other people's office spaces, sometimes to look for materials that might implicate her, but other times just to stand in front of a co-worker's mirror and comb his hair — with the physical and psychological implied threat that "I can go wherever I please whenever I please." This leaves co-workers feeling invaded, much like victims of robberies or assault. A Stork/Cormorant has been known to take pieces of furniture

from other people's offices during non-work hours. There is no logical rationale; discovery is certain and proof not difficult.

The **Spoonbill/Cormorant** lies shamelessly. When finally trapped, s/he seeks to implicate or destroy others. All that matters is that s/he wins or escapes blame. If the challenger is persistent in nailing the Cormorant, s/he will attempt to turn the tables by saying, "Why do you always believe the worst about me?" or "You're always looking to trap me." The truth is that many of the Cormorant's exploits go undetected by those who, indeed, do not look for the worst. So, when the cumulative record is finally partially compiled, the known misdeeds are indeed a tangled web:

Oh, what a tangled web we weave,
When first we practise to deceive!

Sir Walter Scott, *The Lay of the Last Minstrel*

A **Stork/Cormorant** believes that office gossip or universal anecdotes are about him or her specifically. Because s/he is so self-centered, it really never occurs to him or her that a reference may be about a generic issue or someone else entirely.

The **Spoonbill/Cormorant** displays the same self-focus and envy of others that is common to normal Spoonbills, but has stepped over the line by becoming purposely hurtful to others in his or her quest. Sometimes one Cormorant will connive to keep another Cormorant stirred up. That way the conniver gets twice the pleasure vicariously; s/he gets to watch the target get hit and watch someone else get blamed.

The **Stork/Cormorant** uses his or her power specifically to hurt others; s/he really does not care about the issues at hand. The fight has become personal. S/he cannot stand the fact that another has won in a competition of excellence, expertise, or efficiency, so attempts to blast him or her out of the water with bombast and blatantly untrue accusations (that the envied winner/unwitting opponent will spend all sort of precious time and energy combatting). The Cormorant gleefully assesses that s/he has the opponent on the defensive, where s/he wants him. This behavior can belong to an inspector over whom superiors do not exercise sufficient checks and balances.

The **Seagull/Cormorant** purposely uses his or her natural indecisiveness to excess to leave someone dangling. Then, when s/he cannot stand the indecision, s/he makes a decision in the void. The Seagull-turned-Cormorant makes up his or her mind at the most awkward moment possible for others.

The Cormorant divulges information shared with him or her only when it can do the most damage. This could be a White Ibis or any of the other birds gone bad, but it is most hurtful from those that people have trusted with their concerns, mistakes, indiscretions, etc.

The **Green Heron/Cormorant** scholar can use his or her devil's advocate posture and his or her intellectual prowess to hurt others purposely. For example, instead of helping others by posing issues or problems early in the development of a project, or as soon as s/he becomes aware of difficulties, s/he waits to do the greatest damage.

The **Pelican** rarely becomes a Cormorant because s/he is such an optimist and is so focused on delight with life. If you suspect that a Pelican has gone bad, then this is one deeply hurting camper.

The **Great Blue Heron** will tend to lash out defensively to protect himself or herself or young proteges. Rarely in nature or in humans does this happen, and the action is defensive, not offensive.

IMPACT ON OTHERS

People will wonder outloud, "How could s/he have gone so bad in so short a time?" Often there is the insinuation that s/he was victimized. The truth is that s/he got caught and all of those who had born the brunt of his or her schemes have come forward collectively. There is finally power in numbers, whereas each would have been (and still may be) a target individually. S/he has been wreaking havoc for a very long time, often since childhood.

What is the impact of prolonged exposure to a Cormorant? Studies have proved that children who have a Cormorant parent, teacher, or other key adult actually have chemical damage done to their brain because of the trauma and depression experienced at

the hands of their predator. The cumulative denigration takes a heavy toll.

Some churches and theologians are beginning to have a new focus that puts the blame on perpetrators vs. trying to find meaning in victims' suffering. According to Episcopalian priest and theologian, Flora Keshgegian, in *Redeeming Memories: A Theology of Healing and Transformation,* "Most forms of trauma traditionally have been seen as some form of sin, as if the victim were at fault... Well, you're not at fault for trauma... The Christian theological tradition has tended to lump everything that ails the human person under the category of sin." Clarity in this domain is timely, given the widespread revelations of predation by priests and clerics generally.

Prolonged exposure to a Cormorant makes you constantly wary (and stressed). You begin to distrust others sooner, without appropriate cause. The Cormorant's belief that everyone lies (and, therefore, s/he's no different from anyone else) is a cynical take on life. After you have been burned a few times by a Cormorant, you have eaten from the Biblical tree of knowledge; you are no longer the innocent you might have been before. This is usually not an improvement in your quality of life.

HUMOR

Sarcasm is a favorite tool for a Cormorant. Humor is personal. S/he finds a perceived weakness and then hunts, haunts, and taunts. However, s/he cannot take sarcasm from others.

Because s/he is so self-focused, comments will slip out that demonstrate his or her pettiness, selfishness, and general disdain for most others. For example, President Nixon would complain about car arrangements, small details in an official visit, or the fact that the Air Force Strolling Strings were already booked when he wanted them. According to an article by Mike Feinsilber of the Associated Press, "Little Things tended to annoy Nixon," he was sarcastic regarding previous presidents and their guests. He was irritated that he had to do "Mickey Mouse events" and directed that there would be no more than one a day of these (and it had to be scheduled just before noon so that he would not be interrupted otherwise). He sarcastically described the psychiatrists at the

National Institutes of Mental Health as favoring marijuana "because they're probably all on the stuff themselves."

A Cormorant delights in toilet bowl humor. S/he knows that it offends others and it gives him or her a great sense of power that s/he can wallow in it and get away with it. Sometimes it is done with sexual innuendo and the Cormorant thinks s/he's flirting. Storks sometimes wallow as well, so one must observe other characteristics to ensure that one has the right bird.

The mouth of a Cormorant is engaged before the brain (and the brain is not much of a restriction).

POLITICS

> *Great men make small men aware of their smallness. Rancor is one of the unavowed but potent emotions of politics; and one must never forget that the envy of the have-nots can be quite as consuming when the haves have character or intelligence as it is when they have merely material possessions.*
>
> Arthur M. Schlesinger, Jr.,
> *Adventures of the Mind,* "The Decline of Heroes"

In the world of politics, be they small town or national, the Cormorant will wait to see how others will vote before voting and does a lot of wheeling and dealing behind-the-scenes. The Cormorant is a predator vs. competitor and certainly is not one to stand up for a cause in which s/he believes. S/he looks for the highest bidder.

One often sees in the world of politics a unique symbiotic relationship between a Cormorant and a Little Blue Heron on the shallow margin of the stream. The Little Blue is a separate species; he does not grow up to be a Great Blue Heron. The Cormorant (often a staff member or a junior politician) stirs up the water and the debris in the shallows. The Little Blue Heron (a mid-level elected official) picks up what the Cormorant's stirred up, running along the edge of the stream, timing her spurts to follow exactly his movements. The staffer stirs up micro-problems with other politicians or political entities, fires off nasty letters or makes public statements, and gets his Little Blue Heron to support them. The

Little Blue is distinctly led and paced by the Cormorant's agendas and tidbits of information stirred up.

This pattern is also visible at much more senior levels of politics. Candidates for public office must keep a really tight rein on those who work on their campaigns. When elected officials or staffers step over the line into domains of self-interest, they lose the perspective of the moral, philosophical, and legal base from which they should operate. There are far too many examples, from the White House to Capitol Hill and beyond. The old "might makes right" and "I can do no wrong" rationalizations afflict some of our most powerful citizens, and their downfall takes on the proportions of characters in a Greek tragedy. Americans have been disappointed by far too many Cormorants lately.

It is fitting that Richard Nixon rose to the presidency on the wings of Cormorant activity: the modern day witch hunts which found supposed-communists in every venue. It is also fitting that his regime fell because of Cormorant-type activity. But, it took some courageous birds (Storks, Green Herons, and Great Blue Herons) to bring his activities to light and to pursue him relentlessly.

President Nixon is often linked with Machiavelli, whose early study of the use/abuse of power has become a guiding text for Cormorant politicians and tongue-in-cheek Storks. In a New London *Day* editorial, "Still Trickier Dicky," readers learned, "It was a dark, shadowy place — this mind of Richard M. Nixon........as we have known for decades now, there was also Nixon Agonistes, as author Garry Wills called the former president in his book by the same name. Yet, try as we might to comprehend how Richard Nixon became so venal, so petty, so paranoid, our ability to appreciate the full depth of the debasing elements of Nixon's personality still gets tested."

Lies come easily to the Cormorant politician. If trapped, his or her response is akin to Ralph Waldo Emerson's aphorism, "A foolish consistency is the hobgoblin of little minds...." However, Emerson was arguing for continuous assessment of complex issues, not giving an excuse for intentional falsehoods. Emerson also wrote that "liars must have good memories."

FAIR OR FOUL CORMORANTS

There is no such thing as a fair Cormorant!

The foul Cormorant probably has a diagnosable mental illness that needs immediate and protracted therapy. Because there is rarely a sense of responsibility or remorse for his or her actions, true recovery is unlikely. When a pattern of frequent changes in employment emerges, a potential new employer must employ thorough checks of previous places of employment – often difficult under current regulations – to avoid hiring a Cormorant that someone else wants to get rid of!

STRATEGIES FOR INTERACTION WITH A SENIOR CORMORANT

If your boss or supervisor is a Cormorant, transfer to another department or leave the workplace, if you can. If you must stay to accomplish your own goals, then try to avoid interface as much as possible. Get off of his or her radar screen. If you must interact daily, then try to translate the Cormorant's strengths to him or her so that her self-esteem rises enough to make her less combative. However, this will be an ongoing struggle. The Cormorant generally cannot absorb very much positive stroking before being suspicious about your motivation. A compliment triggers an immediate self put-down as a protective device. His or her glass must not approach half full.

A Cormorant enjoys seeing others suffer. If s/he has not gotten her jabs in by noon, s/he will go looking for opportunities to do harm (vs. the Stork who seeks confrontation). The Cormorant's belief pattern is that others do not know what it is to suffer. Any achievements by others are easy and due to luck. And, of course, the Cormorant believes that s/he needs to make his own luck, because life is unfair and tougher for him.

When a new supervisor announces at his or her first staff meeting that "I seem to say things that hurt people's feelings. Tell me...", there is a pretty good chance that you have just inherited a Cormorant. Other birds would not introduce themselves in this way. No one in his or her right mind is going to "tell him."

The Cormorant supervisor enjoys playing mind games with subordinates. For example, at evaluation time, s/he will be verbally caustic. The employee will then be surprised to discover that on the written evaluation which the Cormorant's superiors will see, there is no language like that delivered in private. The Cormorant is cunning and wants to see the junior squirm, but does not want to give his or her superiors ammunition against himself.

The senior Cormorant bullies subordinates, displays temper tantrums, makes sexual comments believing that subordinates will not counter, and may, later, apologize for his or her behavior, saying that s/he was experiencing overload from whatever. However, s/he will still insist on the rectitude of his or her judgment. "I am never wrong. I am always right. You and I operate on different perceptions, and mine's right!"

Her perception is that s/he is the center of the universe. If you have a boss or a parent who cannot seem to carry on a conversation unless it is about her and s/he changes the topic radically, then you have a Cormorant. If you choose to refocus on something directly connected to him or her, s/he will be with you. This may play out even in a scenario in which you are telling your boss that you have a life-threatening disease. The Cormorant will give you one of two replies: 1) "You can't have __X___, because your symptoms do not fit — I looked it up on the Internet." 2) "I don't get it." To which the only retort is, "This is not about <u>you</u>."

The senior Cormorant demands personal loyalty, though s/he is not loyal to his or her subordinates or the mission. S/he tests individuals constantly, setting traps, because catching someone is more stimulating and rewarding than loyalty.

The Cormorant supervisor demands respect positionally and personally — all else is seen as insubordination. The emotional load is so heavy with the Cormorant that rational discussion about the relationship is almost impossible. Once s/he has time to ruminate, s/he will be back sticking pins in the doll again. One strategy is to make the point that you want to keep all of your interaction on a professional level, because you see friendships as non-work relationships. This is a double-edged sword, because

many of us do enjoy our co-workers and relish our daily interchanges. However, the Cormorant superior envies these relationships of which s/he is not a part and will accuse you of favoritism, plotting to undermine his or her agendas, etc.

Like the Chinese, a Cormorant sees an apology as an admission of responsibility or guilt; so, if one attempts to normalize a situation and get on with business, it must be done carefully: "I am sorry that you (the Cormorant) experienced..."

Whenever the Cormorant supervisor must be away, s/he loads up subordinates with tasks, but does not relinquish any authority. Beware of responsibility without the ability to make decisions. Refuse to accept this condition. Upon his or her return, there will be much hind-sighting and you must be prepared to say that you used your best judgment (it will never measure up, but you may have been able to make a difference in an arena that is important to you).

Many authors suggest cutting connections with Cormorants, at least mentally and emotionally if you cannot do so physically. If you worry about them, you give them power. Today's technology allows much of our contact to be telephonic or electronic, with the added benefit of voice mail. That gives you physical distance, time to strategize the most effective response, and gets you out of the face-to-face trap that gives the Cormorant so much pleasure.

Other strategies for diminishing the impact of a Cormorant superior include being very circumspect in sharing personal information about yourself or comments about others. This is the enemy. Do not give him or her ammunition.

Determine what you will do if your Cormorant begins inappropriate behavior. One strategy is to leave the room (with a pre-thought out reason which can be as basic as needing to make a "pit stop"). Another strategy is to be very clear about behavior that you find unacceptable and specify your non-negotiable response. For example, with one particularly nasty relative, a woman declared that she would keep her bags packed so that she could leave on a half hour's notice. (She also kept her plane ticket and car key on her person at all times!)

Document occurrences that seem inappropriate to you. Share them with a confidante, preferably in Human Resources or a similar department outside of your own. Find a mentor of at least the same seniority as your supervisor to coach or advise you and, potentially, protect you should the need arise. Keep a copy of your important papers and notes regarding interface with a Cormorant under lock and key, and be sure that you are the sole owner of that key. A Cormorant has been known to weasel a key to office personnel records, or simply steal it after hours, and remove key materials.

If a senior Cormorant is careless enough to attack you publicly, speak up and let him or her know, as well as all others present, that that is an inappropriate remark/assessment of your work. The Cormorant who has thrown caution to the wind is in the lose-lose mode and the danger level has gone up several notches.

If co-workers come to you with remarks made by your Cormorant supervisor about your independence (translate, lack of loyalty), your arrogance (translate, healthy self-esteem), or your power plays (translate, your visibility to his or her superiors), recognize that your work environment is becoming toxic. If you supervisor's assignment is time-limited, such as in the military, government, or other large corporate settings, you may be able to find a temporary assignment with another department, an internship or corporate leadership program for mid-level personnel, or an opportunity to swap with your counterpart in another location. For example, the Civil Service encourages folks to take hard-to-fill positions overseas with guaranteed return rights at the end of the agreed-upon tour (usually two or three years).

If you decide to leave your workplace, make sure that you document the reasons for your departure with those in echelons above your direct supervisor. They may have been wanting documentation of your Cormorant's dysfunction, but could not actively solicit it without risk of inciting subordinates to complain. Failure on your part to clarify your motivation to depart leaves room for the Cormorant to put his or her own spin on it, and even blame you for things you have not done. Wise seniors require an exit interview with all employees leaving their employ, especially

those who leave "voluntarily." They may have been totally unaware of the maltreatment being perpetrated by your supervisor Cormorant, and may choose to urge you to remain with appropriate safeguards.

In this era of downsizing, employees can feel particularly trapped. They must be innovative to find a way to maintain their career progression and benefits without sacrificing their physical and mental health (and often their performance) to the whims of a malicious supervisor. Glenn Close's character in the recent movie, "The Devil Wears Prada," is a classic Stork-turned Cormorant. There is the hint at the end of the movie that she approves of the spunk of the protagonist who is leaving her employ, but we all sense that she will return to business-as-usual with the next new hire.

Some of the birds are more vulnerable to emotional distress at the hands of a Cormorant than others. Those who are not particularly sensitive interpersonally or intrapersonally, may let the Cormorant's antics roll right off their back. But, it can be excruciating for others who are. Awareness of the existence and dynamics of a Cormorant is step number one. Determining your own personal boundaries is number two. You may actually feel sorry for the Cormorant. Another step in extricating yourself from emotional involvement with the Cormorant is forgiveness, probably not communicated to the Cormorant, but enabling you, the victim, to move on to a new place emotionally and philosophically. The victim does not return to his or her old self after a protracted and/or intense experience with a Cormorant, but rather reaches a new norm that includes more wariness and self-preservation than s/he habitually exercised before. It is part of the price paid for the exposure. Most important is that the victim of a Cormorant not allow the experience to demean him or turn or her into someone s/he does not want to be. That would be giving the Cormorant power.

In the Old Testament, one finds a prayer against the wicked. The "Aha" is the equivalent of today's "gotcha."

1 *...Make haste, O God, to deliver me; make haste to help me, O Lord.*
2 *Let them be ashamed and confounded that seek after my soul: let them be turned backward, and put to confusion, that desire my hurt.*
3 *Let them be turned back for a reward of their shame that say, Aha, aha.*

Psalm 70

When you begin a journey of revenge, start by digging two graves, one for your enemy and one for yourself.

Ancient Proverb

Be careful, lest in fighting the dragon you become the dragon.

Nietzsche, as quoted in
What's So Amazing About Grace?
by Philip Yancey

Forgiveness does not change the past, but it does enlarge the future.

Paul Boese

STRATEGIES FOR INTERACTION WITH A JUNIOR CORMORANT

After a relatively brief period of attempting to correct the behavior of a junior Cormorant, you must begin documenting his or her behavior/performance, because the Equal Employment Opportunity requirements are very specific in this domain. The junior Cormorant, recognizing that s/he is in jeopardy in performance evaluation, will seize the opportunity to muddy the water by accusing his or her supervisor of some fallacious behavior. Most EEO offices will feel compelled to investigate at least the first few allegations from the same individual. If later allegations have the same ingredients, they may be more willing to declare the accusations frivolous. Meanwhile, the supervisor and much of his or her staff are put through an onerous process of investigation which requires non-sharing of the process and isolates each and every person who has been interrogated. It is a demoralizing experience for all involved, except the Cormorant, who relishes every part of it and baits co-workers.

The federal government has a one year probationary period during which a new hire may be released purely because of his or her inappropriate fit in the organization. After this one year, it becomes very difficult to remove an employee for reasons other than poor performance, so it behooves supervisors to be observant during this acculturation period. Educators have similar periods before receiving tenure. Remember that the employee who is dysfunctional early on is highly unlikely to improve when employment is protected.

Spending a great deal of time on minding one's saboteur activities deprives the Cormorant of productive activity. Over time, the Cormorant will fail to measure up to performance requirements, his or her Achilles heel. The one arena in which s/he is mortally vulnerable is performance. Document, document, document.

One particularly treacherous activity by junior Cormorants is to request something in the name of his or her superior which is borderline embarrassing or definitely illegal. When the "gift" is delivered by those seeking favor, the senior is in a very delicate situation. But, s/he must decline the gift and publicly reprimand the junior Cormorant for his or her actions. Otherwise, the Cormorant will continue to sabotage the senior's reputation and the favor seekers will continue to deliver their gifts.

Another subterfuge by a junior Cormorant is purposeful misinformation or omission of information key to the supervisor's successful performance of his or her job. Falsification of data or records is one of the Cormorant's strategies to mask poor performance. The effect of this kind of activity can be life-threatening, depending upon the product, research, and/or service provided.

When a junior Cormorant flaunts the rules, s/he must be corrected immediately, privately at first, and then publicly as appropriate. Co-workers will be relieved that their supervisor notices the behavior and takes action. This strategy also gets part of the aggregate misbehavior out in the open.

When a junior Cormorant begins to avoid contact, beware. Avoidance behaviors include not returning phone calls or e-mails,

turning his or her back, closing the door, and hiding behind a variety of subterfuges. These subterfuges include medical appointments, his or her knowledge of a discrete domain that you cannot possibly understand, and legal/human resource maneuvers. Be sure that your corporate clients are not being hurt in this process.

Another warning signal is a change in style of dress. When someone who typically wears stylish, professional clothing appropriate to his or her original behavior pattern changes radically, take notice.

BOTTOM LINE

Deciphering or decoding a Cormorant is like peeling an onion. With the removal of each successive veil/skin, the core becomes more eye-searing. Only those who know the antidote can emerge unscathed. Most will be tearful. But, a well-cooked onion can be delicious.

> *Don't wrestle with a pig. You both get dirty and the pig likes it!*
>
> Unknown

> *There are people who have an appetite for grief, pleasure is not strong enough and they crave pain, mithridatic stomachs which must be fed on poisoned bread, natures so doomed that no prosperity can soothe their ragged and disheveled desolation. They mis-hear and mis-behold, they suspect and dread. They handle every nettle and ivy in the hedge, and tread on every snake in the meadow.*
>
> Ralph Waldo Emerson,
> "Sobering Realities," *Emerson on Man and God*

> *The common idea that success spoils people by making them vain, egotistic, and self-complacent is erroneous; on the contrary, it makes them, for the most part, humble, tolerant, and kind. Failure makes people cruel and bitter.*
>
> Somerset Maugham as quoted in
> *The Peter Prescription* by Laurence J. Peter

An aggressor is not left unscathed, for it hurts to kill; it hurts to injure others; it hurts to scream and shout profane threats to another. Every time someone strikes another person in anger, hatred or jealousy — whether impulsively or calculatedly — that someone loses something of himself. That person loses his temper, loses his self-respect, loses the commendation of his friends and relatives, loses an opportunity to resolve the conflict and may even lose his freedom.

Martin Luther King, Jr.

As I was walking back from a meeting at the front office, a police car whipped by, came to a screeching halt at my office place a block away, and officers ran into the building, leaving their lights flashing and doors open. In the time it took me to race that block, my mind played through several scenarios. The panic I felt in my whole body underscored that I knew at a very deep level that several employees were capable of behavior that could have occasioned that emergency response. While, as it turned out, none of my scenarios was accurate (a client had entered our office place armed and had refused to leave before accomplishing his non-violent reason for coming), I took formal steps the next day with Human Resources to preclude my worst fears from coming true.

A Cormorant really should never be allowed to rise to a position of power; s/he will destroy a team, unit, or company. The only opponents who are strong enough to combat a Cormorant and win are Storks, Green Herons (if they elect to become involved), and Great Blue Herons. The Seagull can snatch the catch from a Cormorant and alert the Storks, Green and Great Blue Herons to the Cormorant's presence.

Probable Cormorants that I know:

____

____

____

Great Blue Heron

Great Blue Heron

The Great Blue Heron is the largest of the wading birds at Sanibel Island. He can reach about four feet (42 - 52 inches) in height, with a wingspread of more than six feet, but he weighs only five to eight pounds. His body is grey-blue; his neck feathers vary with grey, white, and darker grey; but his head is white with a Navy blue stripe above his golden eyes that trails off in shoulder-length feathers. His dagger-like bill is greyish black and his legs range from yellow-green to blue. Males and females look exactly alike, though the males can be slightly larger (which is hard to tell unless two are together). His feet are webbed to give balance and to prevent sinking into whatever marine muck he is traversing. His feet and the length below his backward-bending ankle (that looks like a knee) give him a spring to help him become airborne rapidly. He flies with his neck folded, as do other herons and his relative, the Snowy Egret.

His eyes serve two different purposes: he can see underwater with a microscopic vision and can see great distances with his telescopic vision. According to Hayward Allen in *The Great Blue Heron,* he has monocular vision (i.e., is capable of focusing each eye independently). So he can have both close and distance vision at the same time. Additionally, because of the "placement of the eyes closer to the back of the head than the tip of the beak, without much movement, the great blue has the ability of 360 degree vision." This comes in handy as a prevention against surprise attack.

He lives in both salt and fresh water environments, ranging from marshes and swamps to ponds, tidal flats and shores. When hunting, he walks very slowly through relatively shallow water or stands absolutely still. While his primary food is fish, he is also a major predator of water snakes. His extensive menu also includes eels, frogs, small turtles, lizards, crabs, shrimp, crawfish, mice, small birds and animals. He does not eat his catch live; instead, he spears his catch and then eats it head first. He can swallow fairly

large prey because his sixth vertebra articulates differently from the others, allowing his gullet latitude. His favorite feeding times are at dawn and at dusk.

While the Green Heron is the only bird to use tools, per se, the Great Blue Heron does use strategies. Much like the Snowy Egret, he will use his feet to stir the muddy water, if he thinks prey is present. Additionally, when he thinks prey is in the water but is not moving in order to elude him, he will dip the tip of his wing in the water, according to Allen, and thus frighten the "stationary" fish into scattering, at which point they become visible and dinner. In a stream, a Great Blue will come up behind fish who are facing the current, eagerly feeding on what the flow brings them. They never see what gets them! Allen also indicates that not only can the Great Blue Heron make a head-first dive from the air into the water, like a Pelican, but he can also arrive feet-first to jump on his prey. Truly a bird for all seasons, he can also swim or dive after fish in water deeper than his usual wading depth.

Unlike the Cormorant who fouls the water in which he feeds, the Great Blue Heron leaves the water rather than foul it. He is the cleanest of the birds from another aspect. He has three sources of powder-down that he rakes with a claw on his foot to release a powder that he can put on areas that have accumulated scum from walking in the water. It serves like most powders in that it absorbs the oils and muck. When he is through preening, the powdered areas are blown free as he ascends in flight.

Although the Great Blue Heron is most often solitary when feeding, he does enjoy a collegial rookery with Cormorants, Spoonbills, and Ibises. Most of the other large birds force-feed their young. However, the Heron will regurgitate his catch into the bottom of the nest so that his young may feed. (Human herons seem to adopt this parenting style.)

His only documented acts of aggression with his lethal bill are instances in which a dog threatened him (James Audubon's dog!) and a man attempted to hold him in order to band him. The attackers lost their lives. In the wild, this aggression usually occurs when creatures have startled him.

BLUE HERON FOLKLORE

Fossil records indicate that the genus to which the Herons, Egrets, and Bitterns belong has been around virtually unchanged for at least 14 million years, according to the *Smithsonian* magazine of April 1999.

The Great Blue Heron appears on an Egyptian mural from 15th Century B.C., according to Hayward Allen. From perhaps about the same period of time, aboriginal American drawings of the heron appear on rock paintings in the Great Lakes region. While the Thunderbird was their sacred symbol of rebirth, the Heron also appears along with drawings of the hunters, their canoes, "the sun and moon, snakes and arrows." Allen ponders the role of the Heron in these men's mind: "... the watchful sentinel... seemed to watch over the family, the grouping. Who would first fly at the need of alarm? The heron. Who would sense strangers even before the human eye or ear could see or hear them? The heron. Who would return, solitary and wise, each year, stay and stand almost as tall as a man? The great blue heron."

Allen further relates that the Great Blue Heron is the symbol used in illustrations of an 1849 treaty of the Ojibwa and Anishinabe: "But to elevate the bird to a ranking of spiritual and cultural importance is to give it the meaning that each of us must feel as we see the solitary great blue standing absolutely still, elegant and imperious, charismatic and silent, a being truly deserving of our attention and reverence."

The Lakota Indians described spring as an eagle, a visionary who can be inattentive to detail. With a wing-spread the size of an eagle's, the Great Blue Heron could fill this metaphorical niche.

The Metropolitan Museum of Art has reproduced a Heron on stationery, adapted from a Greek Carnelian Scaraboid, dating from the second half of the 5th Century B.C.....Notes on the motif indicate that the heron was a favorite household pet in Greece.

In his chapter, "The Decline of Heroes," in *Adventures of the Mind,* Arthur M. Schlesinger, Jr. wrote that Prometheus, "who defied the gods and thus asserted the independence and autonomy

of man against all determinism," was the first hero. His punishment from Zeus was to be chained to a rock with a vulture attacking him. "Ever since, man, like Prometheus, has warred against history... It takes a man of exceptional vision and strength and will — it takes, in short, a hero — to try to wrench history from what lesser men consider its preconceived path... Yet, in the model of Prometheus, man can still hold his own against the gods. Brave men earn the right to shape their own destiny."

Don Quixote and his passionate quest challenged all he met to be their best selves. He put an enormous pressure on them to live up to his expectations, and when they couldn't because they were overwhelmed, they blamed him for their misery. The reigning monarchy did not miss the author's message and imprisoned him for a protracted period. And yet, Cervantes' fictional character, created at the same time that Shakespeare was writing, reached through the centuries to inspire Spaniards living under Franco's regime. They revered the hero who called on their capabilities and their nobler selves. Don Quixote's "unreachable star" has become well known to Americans through "The Man of La Mancha." Robert Frost expresses the same theme in "Take Something Like a Star."

In David McCullough's *John Adams,* we are treated to Abigail Adams, a Heron by all measures. She managed very effectively and efficiently in her husband's absences, keeping the family finances strong by her initiatives and cost savings. She provided endless wisdom and balance to her Stork husband, and to a gifted son (John Quincy Adams) and their other children. Rather publicly, folks attributed John's success to her balance and wisdom. She relished and found strength in her garden and nature, but also enjoyed the social aspects of her various consort duties. She made do and made happy wherever they went. She dressed handsomely, within her limited budget, and carried on an extensive correspondence with the thinkers of her day.

Recently, Elliott M. Offner, a professor at Smith College whose many talents include sculpture, designed the Charis Medal to be worn by faculty members who have served for 25 years at Smith. The description of the medal reads: *The heron, shown on the face, has*

at different times symbolized loyalty, virtuous life, order and good work. Charis, Greek for grace or favor, signified the exchange of mutual benefits and the joy and gratitude attendant on such an exchange.

The group name for Great Blue Herons is a siege — a siege of Herons. The archaic meaning of siege is throne. So, the context for royalty and power are set in the word's history.

OBSERVED BLUE HERON BEHAVIORS

The Great Blue Heron is usually solitary, but will keep the company of other species (though usually not other Great Blue Herons, unless the food supply is abundant). He will keep a fisherman company, waiting patiently to be given a fish (or surreptitiously eyeing the bait bucket). He is unfazed by people on the beach as long as they keep a respectful distance and a rhythmic pace.

He appears at sunrise and remains until well after sunset, fishing by the light of the moon, and is completely untroubled by a fleet of manta rays, barely discernible in the moonlight, cruising the base of the waves right at the shore. At other times, the Great Blue may be encountered on the beach on very dark nights; the only signal that he is there is a shadow or short startled sound.

The Heron waits attentively for just the right opportunity. When feeding frenzies occur that attract most of the other species because fish are trapped by the tides in shallow waters, one rarely sees the Heron — he is usually off in deeper water by himself. His restraint and his stature make him an elegant bird; his sleek lines and striking crown plumage, his patience, and his refined movements make him a distinguished fowl. His camouflage is incredible. When he stands still in front of a tree or bush, he seems to sense his observer's angle of vision and positions himself accordingly. He has no wasted movements. He rarely takes flight, but is elegant when he does so.

While not truly gregarious, he is not ruffled by the arrival of others in his domain — a live and let live approach. He has a collegial, perhaps professorial courtesy; not threatened, he is immediately perceived by others as noteworthy, noble, superior.

He "commands" by his presence, not by control of or combat with others. He seems almost to be a still point on a compass around which others move... some call this "True North."

He does attempt to "save up for tomorrow." A Great Blue Heron was observed by a New England reservoir during a severe drought. As the water became a very narrow creek flowing out of a muddy basin, he caught fish and laid them carefully on a large nearby rock. Unfortunately, an opportunistic raccoon was enjoying his provisions!

His voice is rarely heard, usually when startled or when he is about to fly. For example, a younger Heron came upon a very old Great Blue in the roots of a mangrove tree; at a distance of ten feet or so, the two simply stared at each other, but there were no movements or vocalizations. Then the younger turned and respectfully went back the way he had come.

The silversmith, M. Shields, of Williamsburg, Virginia, makes lovely pins that feature the Great Blue Heron, of whom he writes:

To me, a heron represents some of our culture's most highly cherished values. There is his natural poise and persistence. Completely balanced and quietly energized, he focuses on his tasks, which, from greater distance give him a meditative quality... To me, his true beauty is that he expresses his essential being to us, and we can somehow sense a natural kinship of belonging here together.

OBSERVED HUMAN BEHAVIORS

The Great Blue Heron is the most appropriate leader in a democratic setting. Situationally, there are times when an expeditious leadership is called for, like that provided by a Stork. The Heron's style is slower and messier from some perspectives, but it is more participatory, requires individual responsibility from all involved, and grows others for leadership in the future. Of all of the behavior patterns, s/he places both the greatest challenge and opportunity on those led, on himself or herself, and on those up the chain of command. S/he can learn from others and values their contributions.

The Great Blue Heron sees greys in a black/white continuum and operates on a win-win philosophy. His or her cup is not only full, but overflowing. S/he is creative in problem-solving and in products. S/he can reframe setbacks as opportunities to reposition and excel, and can transmit that optimism to others. Change is his or her middle name, as is ambiguity, but s/he will not procrastinate on making decisions. When s/he determines that s/he can make a good enough assessment, s/he will do so, knowing that forward action is desirable and reformulating is always possible.

S/he is the coalescence of both the traditionally male and female behaviors, the autonomous cognitive activist and the relationship-sensitive connector. By combining the intellectual and the emotional components, the Great Blue Heron models the human ideal. While some see the Heron as ambitious (which would require competing with others), s/he would offer the perspective of "being all that s/he can be" (competing against himself).

In common with the Green Heron, s/he personifies the advice given by Gary Trudeau to students graduating from Smith College: "Ask the impertinent questions, because only by so doing can you find the pertinent answers." The Great Blue Heron will pose those questions with intellectual impertinence, but with diplomacy, so that others can respond appropriately and rise to the occasion. S/he seeks an "elegant solution" — this phrase is generally used by mathematicians and scientists to describe a solution that is beautiful in its efficiency and effectiveness. We need elegance in our workplaces as well.

This propensity for what-if? thinking is aptly symbolized by a caricature of the bird that looks like a stylized question mark — a motif used by Kate Spencer, an artist and entrepreneur in the Caribbean, for her stores' bags. The Blue Heron is a seminal thinker, not one given to sequential, clinical delineation of hypotheses and standard deviations. S/he develops critical questions and intuitive answers and lets others, whose talents better suit close study and analysis, determine the definitive results. Workplaces need all of these talents. And, a Great Blue Heron would be wise to find a Seagull administrator to ensure the orderly

operation of his or her office, because s/he can be overwhelmed by the sheer volume of projects and details in which s/he finds herself involved.

Like the fowl version, the human Heron has one eye on the immediate and one eye on the horizon. Because s/he has a strategic plan, a vision, s/he recognizes opportunities when they come her way. Although these opportunities may not come in the sequential order preferred by some of the other species, s/he sees the connections and is able to make quantum leaps. The human version often admires many heroes, past and present, who combine in a kaleidoscope of possibilities. Harriet Braiker, Ph.D., has warned against women's propensity to use these composites as unattainable models — one human never combined all of these roles and accomplishments. However, the goal is noteworthy, if tempered by reality.

S/he is an extrovert, though not in the same league with the Pelican, because s/he also needs time alone to refuel. Others remark upon his or her sheer physical energy, the volume of creative ideas, and the commitment to the mission. S/he does stand tall in her environment (which is both a bane and a blessing, as we shall see — both a leader and a target). His or her power lies not only in ideas, but in the connections with other empowered people with whom s/he networks. Although it is rare in the natural world to see more than one Heron in a given territory, it does occur when food is abundant. For example, literally hundreds of Great Blue Herons were in the muddy shallows of the Potomac River as our plane landed at National Airport — the nation's capitol has an abundance of niches and needs for the human Herons!

LEADERSHIP STYLE

Unlike the Stork, who is a status seeker, the Great Blue Heron is an achievement seeker. S/he does not seek power, control, and the accoutrements of wealth. Instead, the goal is the accomplishment itself and recognition for it. Whatever a Stork sees that s/he likes or wants, s/he goes after it, with no sense of boundaries or impediments. The Great Blue Heron watches and waits, moving only when an opportunity is within grasp and s/he

recognizes its potential. The Heron eats almost everything, which translates in human terms to being able to thrive in almost any setting in which s/he has the freedom to explore and excel.

The Great Blue Heron is a catalyst. S/he challenges, encourages, and stimulates herself and others to grow as a group and as individuals. S/he is Maslow's self-actuated individual. Risk and forgiveness are nurtured. Much as the bird feeds, the human Heron combines a linear plus an opportunistic approach. This is the key to a period of rapid non-linear change — if all change were in one direction, a linear approach would suffice, but it rarely is.

S/he is able to use effectively the skills and knowledges gained in one domain to enrich another. Unlikely to have a sequential career path, s/he tends to see himself or herself as a "board visionary"... one whose greatest talent is to provide innovative views in a variety of settings. S/he often will remain for only a year or two in any given position, believing that s/he has given her "new eye" and has achieved 80 percent of likely breakthroughs in that period of time; then, on to something new. S/he understands the same in others, supports their productive time while they are in the workplace, and then helps them to move on when they choose. While this creates more chaos, it also creates more opportunity and networks.

S/he enjoys being involved in several different projects at once with the freedom to move among them as s/he is motivated. A classic creative, s/he mulls on one project while moving on to another; then returns to the original when solutions have crept in through her hard-at-work unconscious. His or her work pattern may look undisciplined, and certainly not linear, to birds with other preferences; but it may indeed be the most productive of all of the birds' approaches.

S/he models the willingness to do a variety of tasks in the workplace, often wanting to experience the challenges inherent in each employee's role. But, there is another reason as well. One cannot sense the linkages unless the angles of vision have been experienced. S/he encourages others to relish a variety of interdisciplinary tasks. This not only gives the individuals a

broader perspective of the total organization, which may be fortuitous in upward mobility, but serves the workplace well when flexibility of response to opportunities and challenges can make or break the institution. Trying to create an environment in which employees will share their knowledges, skills, and strategies can only be done when there is no threat in the air, because a natural tendency is to hunker down and protect one's knowledge territory when change is on the horizon.

S/he is a natural leader, collaborator, "servant leader," and teacher who enjoys collegial work with teams and committees, but still needs time alone to replenish. S/he treats youth as young adults and respects their "becoming" status. They feel understood, honored, and empowered. They feel their best selves when they are with the Great Blue Heron. S/he values and recognizes a wide spectrum of people who are given respect until they prove themselves unworthy. This is "grace" in the corporate setting; but the wrath of the Old Testament God awaits those who insist upon doing harm to others (i.e., Cormorants). Capital letter Respect is awarded to those who have earned his or her kudos. Entitlement must be earned, not merely expected. Therefore, it is more relished when granted.

Praise is frequent and sincere. Because s/he thrives upon positive interaction with others whose opinions she values and seeks, s/he is effusive in his or her gratitude for small, but cumulatively important, acts of excellence. There are those in the workplace who see this as manipulative and/or demeaning, but most thrive upon these un-random acts of kindness.

The Great Blue Heron understands and is interested in individuals and groups. S/he relishes diversity of all sorts. In his or her desire to avoid unpleasantness, s/he may let things get out of hand before dealing with them directly. While diplomatic, s/he should not be misunderstood or underestimated as weak... her lethal bill is the consequence. S/he bows to a higher spiritual or moral authority. In fact, those who have encountered a Heron lawyer know his or her compassion and wrath.

S/he is charming, open, optimistic, approachable, and "what-you-see-is-what-you-get." There is an authenticity and consistency

in this behavior pattern; the same person you see today in this situation will be the person you find tomorrow in another scenario. The public and private personas are identical. S/he gives nurture to others and expects nurturing in return. His or her eye contact is direct and focused (micro- and telescopic), but s/he is not confrontational unless severe circumstances warrant. In fact, her face at rest has a smile.

The Great Blue Heron can be thought-provoking (vs. just plain provoking). While s/he can do detailed work, s/he is more excited by the vision of a plan. A "Dennis the Menace" cartoon sums up a Heron's motto: "I color *outside* the lines because there's a whole lot more room there." There is some danger in coloring outside of the lines; s/he has a tendency to get carried away with his or her enthusiasms, talkativeness, and a tendency to exaggerate. S/he does not lie like the Cormorant, but can get trapped outside of the lines.

The Great Blue Heron is a situational leader, capable of varying strategies to fit the scenario and the players, but s/he has a very solid core of constants that permit this flexibility. Just as the bird has a wide array of strategies and resources, the human Heron is both opportunistic and creative. S/he has the sense of being both a spectator and an actor in his or her work, a sensation noted by Anne Morrow Lindbergh.

This duality shows again in the combination of intrapersonal and interpersonal intelligences. The one requires a significant understanding of one's own emotions and motivations; the other interprets the emotions and motivations of others. S/he combines sense and sensibility, a theme explored by Jane Austen. Daniel Goleman's work, *Emotional Intelligence,* and Howard Gardener's *Frames of Mind* explore these personal intelligences as well. The Great Blue Heron stands alone in this combination of intra- and inter- personal insights. S/he truly does want to understand what makes others "tick" so that s/he can be supportive, but this dual capacity and sensitivity are not reciprocated by the other birds. Singular versions appear, but the motivations are not always positive.

A workplace under the leadership of a Great Blue Heron can feel like Camelot. There is an underlying respect for the talents of

each member and a collegiality that allows quantum leaps to occur. Employees in other departments or companies notice the aura and often will seek to transfer into Camelot whenever possible. Or, they will seek mentoring relationships to tie them into the orbit. Those who call Camelot home soon begin to take their environment for granted, not realizing how qualitatively different it is until they interact with their counterparts from other domains. However, the same envy factor that the Great Blue Heron is subjected to on a personal level can also impact the workplace; those threatened by its sheer excellence will seek to tarnish its halo. Even seniors will query the Heron's deputy how it feels to work for someone who always gets the tasks done on time, with accuracy, and with implications noted. This is not an idle question... it may be a search for a flaw which would help to cut the Heron down to a level with everyone else.

BACKLASH AND ANGER

The Great Blue Heron in the workplace feels hurts, slights, and sarcasms deeply and personally. S/he cannot slough these off easily. Unlike the bird itself, who uses a powder-down to clean up after a morning's walk through water and scum, the human version does not easily rid himself or herself of the scum through which s/he has walked. Because s/he knows the slings so personally, s/he would not knowingly hurt others. When others feel diminished by his or her presence (a feeling that they must own, as s/he has not inflicted hurt upon them), they lash out at personally cruel and emotional levels. Puzzled by this onslaught of hate, envy, and discontent, s/he is tempted to wonder what s/he has done to occasion such attacks. Although the temptation exists to toughen up, s/he generally resists, because in so doing s/he would give up what s/he values most. S/he follows Eleanor Roosevelt's advice: "To handle yourself, use your head; to handle others, use your heart."

One of the common interfaces that the Great Blue Heron has is with Cormorants and Egrets who say either, "You think you can do anything!" or "You don't know what it is to suffer. You've gotten everything handed to you so easily." This is the Cormorant's

desire to see others "suffer" and the Egret's annoyance and resentment of the "ease" s/he perceives for others who are less perfect. Herons struggle in ways that often are not visible to others; their challenges are more internal, of a "higher" level (emotional, intellectual, aesthetic, or spiritual), so others do not notice. Their self-discipline is enormous. Ayn Rand addressed this:

All your life you have heard yourself denounced NOT for your faults, but for your greatest virtues. You have been hated not for your mistakes, but for your achievements... You have been called selfish for your courage of acting on your own judgment and bearing sole responsibility for your own life. You have been called arrogant for your independent mind. You have been called cruel for your unyielding integrity. You have been called anti-social for the vision that made you venture upon undiscovered roads....You have been called greedy for the magnificence of your power to create wealth.

The Great Blue Heron must lead his or her life in an exemplary fashion, because s/he is damned if s/he does, and damned if s/he doesn't. If s/he doesn't, critics and envious souls will seek out and share information to do damage. If s/he doesn't, critics and envious others will create misdeeds. Image is the key for this bird, so s/he is vulnerable to the backlash experienced by gifted women, in particular. The key is to know with whom s/he is working and try to meet their needs as appropriate — but the Heron cannot turn him or herself into a pretzel trying to build up everyone else's self-esteem to the magic comfort quotient.

S/he will need to be most careful of Cormorants, for damage control, and be prepared to snatch their catch to establish dominance in his or her own terrain. The Great Blue Heron irritates the dickens out of a Cormorant just by being who s/he is. It is a classic hero/villain clash that is explored in the chapter on the Cormorant. But, if a Cormorant perceives that the Heron is not being brought low by his attack, s/he will double the attack. Sometimes, this Cormorant is not even a person in the Heron's workplace, but one who encounters him or her in a public or official capacity, and lets personal emotions color (black or green) his or her responses. Tales can be told of IRS agents, FBI investigators, and airport security personnel.

S/he seems to provoke envy and backlash from marginal folks; and when they are especially vicious, they attack the Heron's proteges or family members, knowing that this will hurt the Heron even more than a personal assault. Because the Heron does not purposely hurt others, s/he is blindsided when others do so intentionally. If the Cormorant is a junior employee, the Heron may choose to remove him from the workplace or monitor his or her activities very carefully (though this latter solution generally takes more energy than the Cormorant is worth).

Scenarios in which s/he finds herself angry have in common others' failure to factor in the human impact of their actions. The crusader in the Great Blue Heron is activated in circumstances when decisions are insensitive to the very real personal fallout. For example, the mobilization of the same military units in three successive international "policing actions" (so, a back-to-back-to-back impact on families) with little or no lead time. Another example is the officious letter telling residents of a senior congregate living unit that they must meet three days hence to learn the details of reconstruction in their apartments (with the requirement that they vacate their premises for a month). In most cases the repairs could have been optional, but the administration demonstrates callous treatment of tenants in their nineties.

An important note is that many daughters have been socialized in such a way as to dissuade them from expressing anger. While that may have served them well in eras past (though one wonders if it was ever appropriate for their personal safety), today they must learn how to deal with conflict in the workplace appropriately. If they internalize the message that conflict must be avoided at all costs, they are forced to become manipulative to achieve some balance in relationships that are conflicted. This leaves them especially defenseless with Cormorants and other aggressive birds.

The Great Blue Heron's anger is slow to build, but the cumulative effect may be a very quiet bird. This should be a warning to all who are taunting. S/he will deliver, in almost a whisper, a very damaging message. S/he is unwilling to

compromise her ethics, sense of fairness, or appropriate behavior. Those who fail to heed the warning will deal with the Heron's lethal bill.

St. Augustine said that Hope had two daughters: Anger and Courage. The Chicago theologian, Susan Brooks Thistlewaite, has added a third daughter: Joy. The Great Blue Heron seeks a balance in the workplace (and in life) between the short-range strategies of anger (advocacy) and courage, and the long-term strategy of joy. S/he dislikes confrontation, much preferring to be a peacemaker or a diplomat. Some of the other birds see this profile as slippery or spineless, but s/he does not give ground where it matters. S/he negotiates a win-win, if at all possible. S/he does not win every time, but s/he has the courage to try a new tack, looking for an elegant solution. Others can perceive this continuing to seek a more favorable solution as disobedience.

WAYS OF THINKING

Rosabeth Moss Kanter, who has taught at Yale, Harvard, and Brandeis, argued in her seminal work, *Men and Women of the Corporation:*

> *...mental agility is especially essential in times of social transformation.....Trying to lead while the system itself is being reshaped puts a premium on brains: to imagine possibilities outside of conventional categories, to envision actions that cross traditional boundaries, to anticipate repercussions and take advantage of interdependencies, to make new connections or invent new combinations. Those who lack the mental flexibility to think across boundaries will find it harder and harder to hold their own, much less prosper.*

Because the Heron does not see the world in terms of polar opposites, s/he is inclined to believe that there are multiple answers for a given problem. An intuitive thinker, the Great Blue Heron uses both left and right brain thinking, integrating the sequential and the holistic, switching to the mode most fruitful for the current challenge. Neuro-lingual observers can track this movement between the sides of the brain by observing eye and head motion. It gives them clues to thought processes — a fascinating domain.

The Great Blue Heron is both opportunistic and creative. S/he has antennae out and is aware when occasions are fortuitous. But, s/he also has the foresight to start from scratch and build the opportunities that will allow quantum leaps. So, when his or her detractors are busy saying, "S/he was lucky," that is usually a shallow analysis which fails to appreciate the groundwork that had been laid. Remember, this is the bird who strategically fishes upstream behind the fish who are focused on what food might be brought downstream.

Wynton Marsalis, the great jazz musician, had words of wisdom for the graduating class at Connecticut College in 2001: "It is less important to be on time than it is to be in time." In order to be in time, one must have knowledge of the present and a vision for the future, much like the jazz artist who knows the theme and is already thinking about how s/he will add his or her variations when it is his turn. Variety, flexibility, freedom, and the ability to keep pace with the changes allows the musician to take his improvisation as far as his creative juices flow.

S/he is committed to quality work and expects the same from others. S/he also gives and expects consideration in interaction with others. The Great Blue Heron manages by walking around, interacting with staff and with customers. S/he wants to "catch people doing right" — her form of "gotcha" is a hurray! S/he also learns a great deal from using all of her senses in her walk around, ever so much more than from a paper report.

> *Managers are the integrators — the supportive, flexible tissues — that connect beliefs to goals, culture to strategy, performance to reward... They define the overall 'sound' of the enterprise, generate a terrific environment, establish standards of quality, bring in great people, provision creative efforts with needed resources, and establish boundary conditions such as budgets and schedules.*
>
> Max DePree,
> *Leadership Jazz*

VISIBLE SIGNS

The Great Blue Heron's office will be an interesting mix of power photos, plaques, and letters of appreciation alongside symbolic art. Visitors to his or her office will find themselves reading his or her walls in search of what is important to this person. While there will be photos of his or her family, they probably will be tucked into a private space where only the Heron sees them. A good approach is to ask the Heron about a specific piece of art. His or her propensity for the symbolic means that the inquirer will learn something important.

A Great Blue Heron also dresses symbolically and strategically. When dress-down-day, office re-arrangement day, or an off-site planning day comes along, s/he will don the appropriate togs. However, realizing that s/he represents the department both up and across the scale, s/he will usually be in classic clothing with some sort of flair that is uniquely his or her own — whether that is a special tie or lapel pin, s/he is saying something all of the time. His or her choices are never random; they are always purposeful.

S/he is a handsome bird, both in physical stature and presence. S/he usually has some unconscious awareness of the power that comes with his or her profile, but s/he rarely uses it and is generally unaware of the envy it occasions in others.

CONNECTIONS AND CONNECTEDNESS: A PART AND YET APART

Although the Great Blue Heron does a great deal of solitary fishing, s/he is often seen in the company of others. There is a connectedness in the environment a la Dr. Edward Hallowell's comments in *Finding the Heart of the Child.* There is a "sense of accompaniment... a sense that no matter how scary things may become, there is a hand for you in the dark." Those who have walked upon the Sanibel shore at night know that one may encounter a Great Blue Heron fishing or simply walking the shore long after dark. There is in that discovery an amazing sense of camaraderie, in spite of the original startling recognition.

There is a very great desire on the part of human Herons to share what they know, whether that is news within the extended family or within their workplace and community. They see the need for connectedness, because a lot of their food (ideas) comes through the tidbits shared by others. Some of the other birds withhold information as a control issue. Other behavior patterns simply are not interested enough to consider sharing the information (such as Green Herons).

During daylight, one may often see a Great Blue Heron accompanied by a collection of smaller birds, all of whom are consuming what is stirred up by the larger bird. So, the Heron is collegial, if not collaborative.

Corporations have come to value their Herons. In particularly difficult transitions, especially periods of drastic downsizing with lots of union involvement, there are times when a much-beloved giant who has been promoted to corporate headquarters will be given a temporary re-assignment to return until the corporate community and the impacted civilian community get through the particularly rough stages. His or her networks, connections, and ability to inspire others provide the glue necessary for the transition, and then the leadership is carefully handed on to a hand-picked successor who has the expertise and the positive interpersonal skills to take the company through the uncharted territory ahead.

Connectedness to not only those within his or her workplace, but to those within the community in which s/he lives and works (which may encompass a national or international network) is part of the Great Blue Heron's DNA. S/he senses the needs and opportunities, while others may see the energies expended as a short-term waste of time. S/he sees it as an investment that will pay huge and immeasurable dividends. S/he dreams in this scope. S/he not only sees his or her own cup as full, but dreams that those of others will be full as well... not always understanding that each person's vision is really personal and that s/he cannot bring it to pass for them unless they choose to participate in his vision.

It is important to differentiate between being a "peacemaker," which is often the role of a Seagull, and being a "diplomat," the role of the Great Blue Heron. The peacemaker often operates on a win-lose proposition, and is willing to lose himself in order to make peace with a stronger fowl. The diplomat seeks a win-win solution, where both players get their most important needs satisfied and both are pleased with the results.

Frances Hesselbein, former CEO of the Girl Scouts of the USA, now chairman of the Board of Governors of the Peter F. Drucker Foundation for Non-profit Management, was quoted in Sally Helgesen's book, *The Female Advantage:* "The wise leader embraces all those concerned in a circle that surrounds the corporation, the organization, the people, the leadership, and the community."

VIRTUES AND VICES

The difference between a virtue and a vice is often in the eye of the beholder. The issue of pride vs. humility comes up in regard to this bird (and to others such as the Stork). However, it is an issue usually raised by a Snowy Egret. The Great Blue Heron's perspective on humility is that it is not a denial of the self. Rather, it is a choice based on a full awareness and honoring of one's combination of talents, temperament, and achievements. It is a recognition of a healthy pride or esteem in oneself and a choice to give that self in service to others through a mission or calling. S/he gives praise to others abundantly, when deserved, and is more critical of his or her own efforts than any one else. False pride would be taking credit for attributes or achievements for which one has not worked. False pride delights in a sense of superiority (unearned) vs. service.

Rollo May knew that the need for "the courage to create" required one to look fear in the eye and proceed anyway, certain of insights worthy of the risk. Far from being a proud bird, s/he is keenly aware of things yet unknown, challenges ahead, and delight in each "small step for mankind."

Another issue for a Great Blue Heron is loyalty. Storks and others believe that loyalty is hierarchy-driven, while s/he believes

that loyalty is mission-driven: personal allegiance to ethics and values. The Heron typically stands his or her ground on the issue, regardless of a senior Stork's reprimand. The Heron rarely compromises on ethics. The Stork rarely gives ground. The result is usually a draw.

General Dennis J. Reimer, Chief of Staff, U. S. Army (i.e., the senior officer in the Army), was quoted in *The Army Chaplaincy* (1998) as he wrote: "A good leader must have compassion, courage, candor, competence and commitment... we must possess the moral courage to deny this damaging philosophy that says it is worse to report a mistake than it is to make one." He referenced General George C. Marshall on the need for "leaders with the moral courage to tell their superiors when they are wrong."

The Great Blue Heron's loyalty is like that of the shepherd who seeks the one lost sheep when 99 of his flock are safe. S/he is also like the shepherds referenced by Jesus in the New Testament. Whereas shepherds in most areas of the world are at the back of their flock, watching for aberrant behavior and then prodding as needed, the shepherds Jesus knew led their flocks, showing the way, protecting against what might be ahead.

HUMOR

The Great Blue Heron enjoys irony, light satire, and good hearted humor directed at human foibles. S/he enjoys political satire, such as that offered by "The Capitol Steps," a group of gifted performers who gently mock the current powers-that-be in irreverent costumes, comments, and comic renditions of well-known songs. However, when the humor becomes sarcastic, hurtful, or tawdry, the Heron adopts his or her watchful stance. Be prepared for his departure or his lethal bill.

Humor is found in the appreciation of the daily, the delight in the collisions of desires and disillusionments, and puns of all sorts. Literary or symbolic allusions provide instruction and amusement. Repartee with others makes for a good verbal/intellectual fencing match, but all within the constraints of good taste.

The Great Blue Heron is more likely to tweak and tease those

whom s/he admires and enjoys than to use his or her humor as a weapon. S/he will also make light of his or her own shortcomings. Laughter will be heard throughout the workplace as people are at ease; even in tough mission areas, they know the healing power of being amused by oneself and others.

Diane Sawyer's advice in *Working Woman:* "I just want the generation coming up to laugh a lot along the way. Up or down, love the ride."

POLITICS

The Great Blue Heron has a superb sense of what others need and want. President Reagan is often seen as a good example of the improvisational performer (who used note cards for critical speeches), but had an uncanny sense of the ripple effect. He saw analogies in all that he did, often citing the experiences of individual Americans to underscore his point. His theatrical flair ennobled events (such as his State of the Union messages and his painful reunion with Beirut Marines and their families/widows). Conversely, the Spoonbill uses the theatrical for soap opera flavor. The Heron can feel empathy with wounded veterans, whereas the Spoonbill evokes pathos.

The Great Blue Heron can serve as a catalyst. His or her "divine discontent" (à la Ralph Waldo Emerson and Coretta Scott King) prompts positive drive and growth. The Heron's nutshell communiques are really marketing slogans. The most effective have been President Clinton's "Bridge to the 21st Century" and President John F. Kennedy's "Ask not what your country can do for you, but what you can do for your country."

One especially grave danger for a Heron politician is that his or her exploration of the nuances or the middle ground of issues vs. the polar extremes may be seen by others as two-faced, deceptive, and unreliable. Particularly in this age of sound-bytes, this politician is not given the time or space to explain the complexities that many of our major decisions require. While Stork politicians tend to have a very blunt speaking style, the speaking style of a Great Blue Heron, such as General Colin Powell, is described by *USA Today* as "soft around the edges" — a description of the

diplomacy needed by an effective Secretary of State.

Aside from those who have made politics their life's work, one additionally finds a Great Blue Heron who has spent his or her career in a field such as the military, space exploration, medicine, or education. Because s/he is such a giant in his or her arena of lifetime work with unquestionable integrity, and s/he does not have the political baggage of those who have labored in the vineyards, this apolitical creature may actually be wooed locally or nationally by both major parties.

Maybe we Americans should start worrying as our so-called individualist society develops a cult of the group... are not these group tactics essentially means by which individuals hedge their bets and distribute their responsibilities? And do they not nearly always result in the dilution of insight and the triumph of mish-mash? If we are to survive, we must have ideas, vision, courage. These things are rarely produced by committees. Everything that matters in our intellectual and moral life begins with an individual confronting his own mind and conscience in a room by himself.

Arthur M. Schlesinger, Jr.,
Adventures of the Mind, "The Decline of Heroes"

FAIR BLUE HERON

Because a Fair Blue Heron thrives in the upper stratosphere, s/he often does not pay attention to the vagaries and vicissitudes of folks who seek to do her ill, or to her spouse or workplace deputy who may feel unappreciated as s/he focuses on loftier issues. A Fair Blue Heron has a focus, stamina, and self-sufficiency that may cause others to question his or her connectedness and commitment to them. As s/he focuses on work or personal projects, s/he may seem to be in another world. Others generally relish having his or her attention and resent the time that s/he needs alone to replenish intellectual, emotional, and physical energies.

S/he is charismatic. While that is a much-overworked word these days, this bird has both personal and corporate charm. His or her agility with ideas and words and the tactfulness with which s/he shares them are magical.

The Great Blue Heron enjoys acquiring beautiful things. This acquisitiveness is more an aesthetic delight than the greed that motivates some of the other birds. S/he enjoys lovely art, music, furniture and colorful surroundings. But, s/he is also generous in sharing them.

Thomas Jefferson, the incredible dilettante who achieved remarkably in almost every domain in which he dabbled, was a fair fowl dipping into a foul fowl. As he became excessively acquisitive, he became profligate financially. For example, short of funds, he sold all of his books to the new government to establish The Library of Congress. As soon as his income improved, he began to buy again. Although disciplined intellectually, his inquisitive nature led to endless changes at Monticello and his acquisitive nature led to financial ruin. And yet, he penned some of the most powerful political tracts known to man and had the foresight to commission Lewis and Clark's expedition. He alone vastly increased Americans' knowledge and achievement in such disparate domains as horticulture and architecture. His idealism, optimism, and vision led others to new levels as well.

He, like John Adams (a Stork), was oblivious to the Cormorants operating around him. He trusted folks like James Madison, who actually blocked a letter which Jefferson wrote to Adams praising him. Madison believed Jefferson could be hurt politically by expressions of friendship, according to David McCullough. So, the two giants never knew that an admiring message had been intercepted and they remained at odds with each other until their deaths. Storks and Great Blue Herons, two very powerful leaders, must communicate directly with each other in order to preclude others' manipulation.

In *John Adams,* McCullough compared Adams and Jefferson: "It was Jefferson's graciousness that was so appealing. He was never blunt or assertive, as Adams could be, but subtle, serene by all appearances, always polite, soft-spoken, and diplomatic, if somewhat remote... With Jefferson there was nearly always a slight air of ambiguity." He also noted that Jefferson's personal philosophy was to get through life with the least pain possible. "He shunned even verbal conflict."

FOUL BLUE HERON

A Great Blue Heron can risk too much in his or her enthusiasm and desire to achieve. S/he may become excessive, take short cuts, and become overwhelmed. A candidate for burnout, because of his or her dual commitment to mission and to people, s/he puts other duties before her own well-being. Once burned out, s/he may take a long time to recover a healthy balance.

As a Heron becomes dysfunctional, s/he fails to collaborate and share tasks, and tries to do everything him or herself. Because of his or her enormous energy, ability, and perseverance, s/he manages for awhile, but burnout usually occurs. This may take the form of depression (more common in women) or physical illness due to an impaired immune system (resulting from physical overwork and emotional hemorrhage).

S/he needs to learn how to set limits for himself and learn how to say "No" or "Not now" to those who do not appreciate that his or her energies are finite. Otherwise, s/he will begin to lash out at others and they will be puzzled by this uncharacteristic response. His or her bill is lethal when s/he is startled or threatened.

A Great Blue Heron who has been raised or managed by Cormorants, Storks, and/or Snowy Egrets may have experienced so much attempted external control that s/he feels robbed of independence and a sense of self-achievement. The resulting guardedness against any dependence on others may take a long time to heal, so that his or her natural expansiveness may return.

STRATEGIES FOR INTERACTION WITH A SENIOR BLUE HERON

> *Many will see themselves as hard workers, no matter how they are treated. They will give their hands and, even, their brains to a boss. But they will give their hearts only to a leader, and the feeling we experience when that happens is something a boss will never know.*
>
> William Glasser, M.D.,
> *Choice Theory: A New Psychology of Personal Freedom*

Your Heron supervisor will delight in your initiatives. S/he sees his or her role as "strengthening creativity, confidence, and competence" in herself and others — the prescription for leadership, according to Laurence Peter, of *The Peter Principle.* S/he relishes quantity and quality in ideas, knowing that through a sheer volume of what-ifs, the workplace will excel in forward thinking. S/he has a tolerance and a taste for productive risk taking. "Let's try it," will be a favorite response. Many proposals really do not require monetary commitments; often options simply require attitudinal change and/or a rearrangement of personnel and production assets. These come under the category of doing more... often with less.

S/he will, under most circumstances, "read" you well and respond supportively. However, you can help him or her know what it is that you need, thus expediting the process, because if s/he is burning his or her candle at all ends and in the middle, which is classic behavior, s/he may not take the time to listen carefully, and will feel badly that s/he failed to do so.

S/he expects quality work from herself and others. Give him or her no less. Do not expect her to take kindly to editing your work, which you could have done with spell check on your computer. When s/he has given you guidance in a specific domain, continue to follow it in all succeeding instances where it applies. S/he does not suffer fools gladly. Visions are too glorious, time is too precious, and energies too limited to revisit the same terrain.

You will rarely experience anger from your Heron supervisor... more a grave disappointment that someone s/he has empowered through collegial leadership has abused his or her freedom through irresponsibility or a lack of initiative. When this happens, s/he will reluctantly move to a White Ibis, paternal style of leadership that is highly structured and more restrictive — until s/he feels comfortable with an employee's reliability.

S/he is a marketer par excellence. As you listen to him or her, try to give your constructive criticism to what seems substantial and what seems more shallow. The Great Blue Heron can get on a roll and fail to differentiate between a solid infrastructure and a

facade. S/he can become impractical as s/he gets swept up in his or her enthusiasms.

S/he does not relish being embarrassed. Do not blindside your Heron. Much like the wild creature, s/he will respond with his or her lethal bill. Advise, forewarn, discuss. If you are leery of some aspects of a current project, explain your reservations and have a different solution to offer. S/he will value not only your regularly scheduled supervisory time, but will accommodate a short-notice request to discuss an emergent issue. Aware that some folks need just a few moments' time to calibrate their current project and keep the momentum going, s/he will make time as soon as s/he can. S/he, too, gets deep pleasure from championing an exciting venture.

Some find a Great Blue Heron hard to "read." Because s/he is flexible and willing to listen to others' input, s/he may be confusing to those who need more structure to be productive. Let him or her know that you need specific guidance on your project if s/he leaves too much of it up to you for you to feel comfortable. It is part of his or her style to allow room for personal growth, but s/he would not want that to translate to personal discomfort.

S/he models the behavior s/he values. S/he arrives on time for a meeting, respecting the time commitments of others, has notes on items s/he wants to address, but always allows time for you to speak about what is on your mind. Plan to have something on your mind! Because s/he is a coach at heart, s/he wants you to initiate ideas and take the responsibility for fleshing them out, once you both have discussed the possibilities and implications. S/he sees his or her role to be growing young talent to promote, because s/he may be off on another quest and will need an able successor.

S/he will want you to have the credit for what you have accomplished. His or her reward is having nourished your accomplishment. S/he will expect you to "Pass It On!" — to enable others in the same way that you have been helped, because the likelihood of payback is so rare (and, frankly, so crass).

The Great Blue Heron does keep long hours (often starting before dawn and continuing after dusk). But, s/he does not expect,

nor often want, company from employees in these extended periods. Because s/he has been in their company all day, s/he needs time to refuel, to think longer thoughts. Often a Heron will do this at home vs. in the office. S/he is always thinking, looking for the creative links, and because her antennae are out, s/he will find far more than most of the other birds. S/he will want employees to work their normal hours and keep a healthy balance in their lives, so that when they are on the job, they can be truly on the job.

The Great Blue Heron values both individual achievement and team productivity. S/he sees a need for both approaches, selecting the most appropriate for given tasks and given people. S/he recognizes the strengths of each of the behavior patterns and tries to assign work accordingly. Professional development is also highly valued. If you know of an opportunity that you think would add significantly to your skills and strategies, speak up. S/he will attempt to enable a fair dispersion of opportunities to all employees in the workplace and appreciates your research and initiative.

STRATEGIES FOR INTERACTION WITH A JUNIOR BLUE HERON

Productive interface with a junior Great Blue Heron requires a realistic assessment of the capabilities of this person. If s/he demonstrates innate intelligence, a fervor for the mission, and a great desire to perform up to your expectations, then become a coach and enjoy the ride. Experiences with college-age mentees in bureaucracies have demonstrated the two-way excitement that occurs when professionals are allowed to share what they know (which amazes their young admirers), and young people are able to remind their mentors what motivated them to begin their career journey. There is also the satisfaction that one is growing a knowledgeable successor so that one's life work will be continued.

A junior Great Blue Heron requires support, direction, and appreciation, but not micro-management. In fact, if you operate with an open-door policy, s/he may overwhelm you with ideas. It is important to let him or her run with the ones that you think may

have value. There may also be an easy way to let him or her explore the ones that are questionable; s/he may learn quickly that those need to go on the back burner. But, find a way not to turn off the enthusiasms. Because this bird generates so many what-if options, s/he accomplishes far more than most other behavior patterns, just by the sheer number of explorations (and failures). If the climate permits questionable suggestions, you may get quantum leaps far beyond the normal ratio.

If you must turn down a Heron's suggestions, do so gently and with careful explanations. That will allow your Heron to be strategic in future proposals, but it will not turn off his or her creativity. This is the hen who laid the golden egg... keep her producing!

S/he may need protection in the workplace. A mentor would be a good idea as your junior star often evokes envy and hatred from others, often with no clue as to its existence, and for sure, a puzzlement as to its cause. His or her sheer presence reminds others of the expansive horizons for human beings. When one's own horizon seems less glorious, the golden girl (or guy) exacerbates the already low sense of self-esteem... usually through no conscious effort of her own.

BOTTOM LINE

> *... leaders of the future will need the traits and capabilities of leaders throughout history: an eye for change and a steadying hand to provide both vision and reassurance that change can be mastered, a voice that articulates the will of the group and shapes it to constructive ends, and an ability to inspire by force of personality while making others feel empowered to increase and use their own abilities...*
>
> Rosabeth Moss Kanter,
> *Men and Women of the Corporation*
> *Chapter 9: "World-Class Leaders: The Power of Partnering"*

... Your playing small doesn't serve the world. There is nothing enlightened about shrinking so that other people won't feel insecure around you. We were born to make manifest the glory of God within us. It is not just in some of us; it's in everyone. And, as we let our own light shine, we unconsciously give other people permission to do the same...

Nelson Mandela

"The appearance of a great man," wrote Emerson, "draws a new circle outside of our largest orbit and surprises and commands us."... Great men enable us to rise to our own highest potentialities. They nerve lesser men to disregard the world and trust to their own deepest instinct.

Arthur M. Schlesinger, Jr.,
Adventures of the Mind,
"The Decline of Heroes"

Probable Blue Herons that I know:

____

____

____

____

Scurry Birds & Skimmers

Scurry Birds and Skimmers

Our Scurry Birds are technically a collection of Plovers and Sandpipers. They are very difficult to identify by species or gender, so I have clumped them all together because of their general size and behaviors. Small, with fluffy speckled feathers and some rings around the eyes or neck, these dear little birds when young are at risk from a variety of predators, including the Seagull.

They are small waders that generally feed and travel in flocks; but, they can be alone or with only one other Scurry Bird. They arrive on the beach at sunrise and usually depart at sunset, with seemingly very few restful moments in between. However, as noted in "Tracks in the Sand," they enjoy resting in human footprints. Each chooses his own footprint and all face into the wind. When not in motion, they are incredibly camouflaged by the sand and detritus on the beach.

Most of the time they look like a brown furry wave on legs, as they weave around the incoming and outgoing waves, continuously poking their heads into the wet sand, looking for invisible morsels of insects, worms, and mollusks. They always seem to find something worth consuming. Due to the energy expended for each piece of food, they seem to be in constant motion. They look like the old-fashioned child's toy that had a ring of chickens that would peck as the ball on a string beneath the paddle swung around in circles.

Although their pace seems frenetic compared to the other species, their strides are measured and rhythmic. They move around humans on the beach like a scurrying shadow, but do not take flight unless frightened by sudden movements or feel trapped between groups of people approaching from two different directions. If one bird gets anxious and heads away from the water's edge, the whole flock follows. However, there does not seem to be a leader, per se.

Although they arrive as a species-specific flock, they intermix with other shore waders such as the White Ibis. On the mud flats,

they are also in the company of Roseate Spoonbills and Snowy Egrets. They accommodate injured individuals in their flocks, with the speed of the feeding being designed to keep the handicapped protected. The White Ibis is the only other species seen with injured members in the flock.

The smaller the bird, i.e., Sanderlings vs. Plovers, the more likely they are to avoid the surf. Leg length and bill length are visible indicators of the depth of water in which they feed and the depth to which they probe the sand and mud. Constantly chittering, they provide a mellow vocalization disproportionate to their size. They are gregarious during most of the year, but breed as solitary pairs (just the opposite from the bigger birds).

OBSERVED HUMAN BEHAVIORS OF SCURRY BIRDS

Because the Scurry Birds are so prolific on the beach and in our workplaces, it is important to recognize them. Unlike the fowl versions, who remain true to their species type throughout their life, the human version usually "grows up" into one of the other behavior patterns, such as a Seagull, an Egret, an Ibis, or a Spoonbill. Ever accommodating, they scurry to get minor tasks done, but try to stay out of the way of most of the larger birds. As soon as they have passed by the larger bird or the human, they continue their tasks as if they had never been interrupted in any way. A supervisor knows this all too well when s/he has tried to impact their way of doing a task! However, because they can be so accommodating, they can run the risk of being pawns in the service of ill-intentioned behavior patterns.

They exhibit a great deal of activity, but do not necessarily accomplish much each day. They believe that by their incessant industry and energy that they will look as if they are important to the team. They depend upon wave action to bring them food. They do a lot of little things that keep the workplace clean, balanced environmentally, and they sometimes rest in larger footprints. Of particular note is that the tail of some of the Scurry Bird species flips up and down in a rocking motion, even when they are standing still. They give the impression of constant motion.

Donna and Lynn Brooks write in *Seven Secrets of Successful Women:*

While I was waiting for a senior executive a while back, I became aware of a group of people gathering for a meeting behind me. On the hardwood floor, I could hear some people clickety-clacking at a very fast pace and chatting with others. Then there were others en route to the meeting who were moving and speaking more deliberately.....Sure enough, all the ones scurrying to the meeting appeared to be frazzled lower-level employees, while the others looked like senior management. Even if it was only the image they portrayed (I didn't actually know who they were), there surely seemed to be a fairly clear-cut difference in rank.

The Scurry Birds seem to be more of a DEVELOPMENTAL STAGE than a BEHAVIOR PATTERN. Most Scurry Birds become a more clearly defined behavior pattern over time in their workplace, becoming easier to identify and, therefore, motivate. However, there are a few who remain Scurry Birds forever.

In these times of corporate downsizing and the rise of truly small businesses, the niches for the Scurry Birds may be disappearing. Few workplaces can support their modest production. The Seagull, who often occupies a position as a senior administrative assistant, dislikes having to spend time supervising when s/he could simply get their work done expeditiously himself or herself. Even in military settings, the unrated junior enlisted member must show initiative and strike for a specialty rating, or s/he is seen as unproductive and risks early release.

Skimmers

Skimmers occupy a very unusual niche in the fowl environment. They resemble the Royal Tern, a very handsome bird with gray and white body feathers, a black cap and black tail feathers, and a prominent orange bill. However, the Skimmer, who often shares the beach with the Royal Terns, just to confuse observers, has a black body, with white below his eyes and under his bill. His bill is half orange and half black, and the upper lip (mandible) is shorter than the lower lip. Also, he has orange legs, if one ever gets a look at them.

While bird books list him as "unmistakable," he is very mistakable for his beach mate until he takes to the air. Then, his remarkable choreography and acrobatics make him one of the most entertaining and beautiful creatures in flight. Artists have enjoyed showing several of these awe-inspiring birds in formation; they invite graphic delight.

Looking like Mother Nature's Blue Angels, they swoop, loop, and zoom past the water's surface just close enough to trail their lower mandible like a sieve through the water. If anything edible is in their path, typically small fish, their bill snaps shut and they swallow their catch. Because they do not need to be able to see their prey, they are most active just at daybreak and evenfall. Sometimes, a solitary Skimmer will swoop down a pond at the Bailey Tract on Sanibel at dusk when small fish rise to the surface to feed. If it is your first sighting, you will wonder what you have just seen. Early morning birders will find a small flock wherever there has been recent dredge material put down to counter erosion. Skimmers have a fondness for recently disturbed environments and will move on as the natural grasses stabilize the beach.

This is the epitome of the shallow feeder. His trajectory accesses only the depth of a portion of his bill making a knife's edge through the uppermost layer of water. The conditions in which he is most effective are those of very shallow waters due to exceptionally low tides. Because food is then concentrated into shallow pools and streams, his superficial sieve catches relatively more food than when he swoops over deeper water. Sometimes in the shallow water scenario, he is accompanied by a trail of other species, who cannot mimic his swoop, but sense that there will be food below what he is catching.

Charley Harper, in his wonderfully whimsical book of drawings and descriptions, *Beguiled by the Wild,* shows a tier of five Skimmers fishing, the rays of the moon mimicking the lines on their feathers:

> *Behold how black skimmers fish for their finny food. As the tide falls and the moon rises, a squadron of skimmers shimmers over the glassy cove, shallow-plowing the shallows with their razor-thin*

lower mandibles, scooping up minnows by the many from the mini furrows. It's a little like seining with a string, and you have to wonder if they don't come up with a lot of flotsam and jetsam.

OBSERVED HUMAN BEHAVIORS OF SKIMMERS

The Skimmers have a great resemblance to the Pelicans in their human form. However, the Skimmers feed only at dawn and dusk, and feed only with the shallowest technique. They literally skim across a still surface (no agitated waters for these folks) and hope that they will encounter tasty morsels. Awkward on land, they are beautiful in the air.

The Pelican, by contrast, dives headlong into the water after having spotted his meal. He enjoys life to the fullest, but he knows that he is fishing in fertile waters. The Skimmer is fishing blind. He chooses not to fish in turbulent waters, but does skim over areas that have recently had dredge material dumped (and therefore, there may be a release of food materials that were on the bottom of the dredge site before being vacuumed up).

The Skimmer rests and nests temporarily on these dredge material sites. But, he moves on quickly when the site has stabilized.

Skimmers are fairly rare in the workplace, but it is important to know that the species exists. They may indeed be the folks who specialize in temporary work, regardless of their specialty. No longer the domain purely of administrative and secretarial substitutes, agencies tout their professional workers as well. Interim positions may attract those who relish the shallow connections that are typical after the departure of a senior professional, often "under a cloud" — the parish or corporate equivalent of a dredge site.

A classic human example of a Skimmer was spotted in the supermarket at Sanibel Island. She was a pretty, older woman whose hair with a headband and classic clothes resembled the Ivy League college girl she probably had been. So focused on the few

items that she wanted to buy for her party, she would leave her shopping basket across the head of an aisle so that no one could pass by while she was off searching. When she would return with an item, she would change her mind. She was not obnoxious or even conscious of her actions. No eye contact occurred; she was totally oblivious and totally wrapped up in her little menu.

Unlike the Pelican, who plunges into life and bobs on the surface afterwards, the Skimmer has developed a truly remarkable approach to life. For example, a young British naval officer's wife, when asked how she coped with moving so often and having children away in school so much of the year, explained that she truly enjoyed visiting new parts of the world and meeting lots of people. She said that she didn't get very attached to people, places, or things, so it was easy to move on. A day or so later, she opened the conversation again, noting that as she had sat in the sauna, it occurred to her that what she had said might seem terribly shallow....... To her credit, in a Pelican kind of gesture, she offered to invite a number of her compatriots for tea in order to further explore the coping strategies used by military wives.

Like this young wife, a Skimmer may, under certain circumstances, mature into a Pelican. You know that you have a Skimmer when you simply cannot "put your arms around" this individual. S/he is so ephemeral, shining, attractive, gossamer, that even after being acquainted for awhile, you still feel at a loss to name his or her interests, hobbies, or passions. You don't know if s/he agrees with you or not. There's no "there" there. S/he is not offensive, just elusive, illusionary. S/he dips ever so little into life visibly — s/he may be putting on a protective facade, if s/he belongs to one of the other behavior patterns, or s/he may not have learned how to deeply enjoy life as a Pelican. Skimmers barely ripple the water they pass through.

Probable Scurry Birds or Skimmers that I know:

____

____

____

____

Pulpit Feathers

Pulpit Feathers: Pastoral Versions of the Birds

Sadly, in this day and age (and probably in all ages past), we have far too many instances of ministers and priests "gone bad." While we recognize that they are human, they are professionals with a calling that requires the highest level of behavior – just like doctors, teachers, counselors, and many others. We *entrust* them (a wonderful word), with our moral/spiritual confidence and guidance. When they abrogate that trust, we feel betrayed – betrayed on levels that range from the physical to spiritual.

We encounter our pastors and priests in a variety of settings. In many instances, we are able to choose where, when, and if we worship. In other instances, we are given no choice. Our parish is "assigned."

In Protestant denominations where the national or regional hierarchy determines whom to send to a given church and when, parishioners still have parish-relations committees to guide and direct affairs within their own church. They can determine if someone is a "pass-along," someone the hierarchy hasn't the courage to remove from service for any number of reasons ranging from ineptitude and incompetence to non-compliance with church policies; and they can exercise "caution/refusal" pre-emptives. Individual parishioners also have the choice to attend and contribute elsewhere. Their loudest vote is with their feet!

In Protestant denominations which exercise the right to call the minister of their choice, the following profiles will be very helpful in the search process. Remember that the Cormorant is one of the birds gone bad. A careful check back through several previous churches will be very important. The most recent/current church may be very eager to get rid of this particular individual and may be less than forthcoming with the gaining church. The current church may also have made "the agreement from Hell" with the outgoing minister that it will not speak against him/her in his launch to find a new church. So, the search committee should converse with at least one or two prior churches. They will be free to be honest about the contributions or lack thereof of the potential candidate.

When Protestant denominations assign a search committee with the task of finding a new pastor or associate pastor, they often provide processes that enable the individual church to define its own demographic/philosophical profile and a position description for the pastoral candidate. Regional or state entities may also advertise and funnel candidates to the search committee, resulting in more than one hundred candidates' profiles. How do they begin to sort through them all?

First of all, the search committee should determine the bird profile(s) they find most appropriate for their setting. The following ministerial birds' profiles will help determine "who's who" in the cast of applicants.

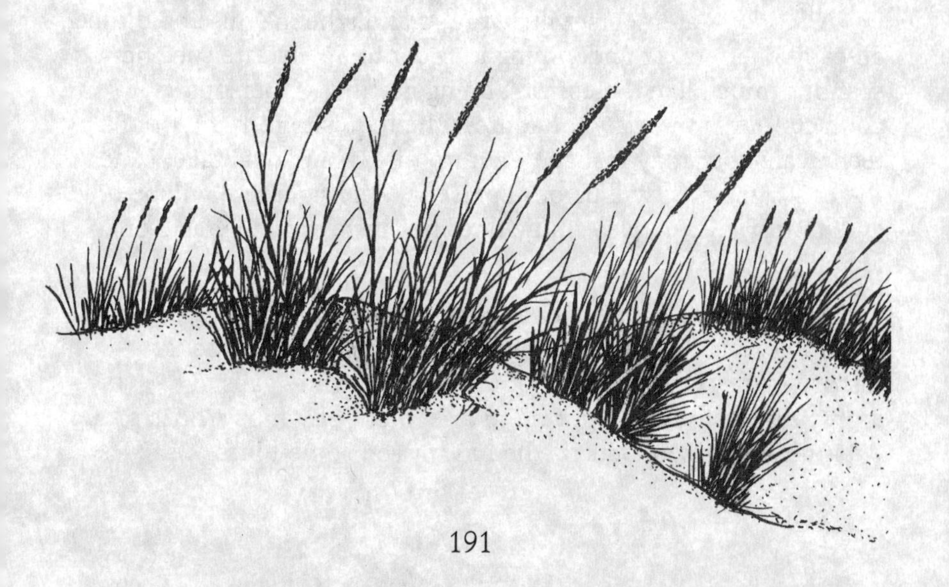

Wood Stork: Pastoral Version

Hard as this is to believe, a minister operating from this position will interpret the Lord's Prayer as a demand in the imperative mode: "Give us this day our daily bread." Most others interpret this as a request. S/he will select a religious environment in which s/he has absolute sway. S/he knows what is best for moral values, church budgets, program protocols, and decisions of all sorts regarding the parish and its facilities.

STORK RESUME STYLE

In order to help you sort through applicants for positions from whom you receive resumes, you may notice some of the following characteristics from a Stork. The writing will be in absolutes. A Stork will document large accomplishments, often exaggerated, and will take credit for the entire results instead of crediting team efforts. The actual resume or form may be messy or have typographical errors. A Stork is not given to editing, which s/he sees as menial (though some in higher positions have come to understand the importance of accuracy of facts and language for which they will be held responsible).

A Stork's statements will be broad and s/he may use bold type, all-capital letters, or underlining in order to accentuate specific sections (much like the use of jabbing gestures in personal contacts). A Stork may not address the requisite form questions technically, but may reshape them to his or her advantage. And the ego-centric "I" will be omnipresent. The signature will be bold and almost unreadable — s/he does not have time to waste on legibility.

A Stork has a propensity to exaggerate successes and blank out on problematic sequences when relating his or her personal history. S/he will glide over or totally ignore any gaps in employment, moving right on to a segment of which s/he is proud. An interviewer wonders if s/he has missed something.

If asked to comment on a personal weakness, the Stork may indicate that s/he is a workaholic. The written rationale will be that s/he is so committed to the mission that long hours are a small sacrifice. The truth is that s/he is ambitious, wants to be available whenever the boss is still in the office, and may be committed to the mission because of the sense of power received by being connected to it. Some Storks like to avoid family dinner hour and enjoy the personal attention they receive from their spouse when they arrive home late. Other bird types have different reasons for this increasingly common phenomenon.

STORK SERMON STYLE

If you find yourself on a pastoral search committee (or any equivalent group selecting a corporate lawyer, new professor, college president, or non-profit high-profile director), the following formal speech profile may be very useful.

A Stork punctuates words with an almost percussive quality, either by use of his or her hands or head. One religious leader actually stands up on his toes, whether speaking to a religious assembly or not, and thereby uses his entire body as an exclamation point. S/he has a "sermon voice" which is almost theatrical versus a natural, familiar tone. A Stork will talk quickly when s/he talks "naturally," indicating his or her rush to deliver information as opposed to observing or listening to others as s/he interacts with them. Or, if the theatrical mode prevails, there will be a holier-than-thou demeanor that may slow the speech pattern so that the peons can understand.

A Stork uses aggressive body language such as choppy hand motions and finger-poking or pointing at "you." His or her hand movements may be close to the body; they certainly are not lyrical , open, and inclusive. Anger may peek or leak through.

A Stork uses "I" a great deal, but may also use the "royal We." S/he rarely includes himself or herself with the congregation in a truly collaborative "we," but speaks to the congregation as "you" (the sinners). The Stork rarely comes out into the pews during the worship service, because that is the level of the congregation. If it does happen, during a baptism, for example, the congregation may feel invaded or manipulated.

If there is a lexicon, a recommended scripture for the week, the Stork will use it instead of choosing his or her own scripture. The interpretation will tend not to be contemporary or daily-life related. Historical examples will prevail. His or her preference is for the Old Testament God versus the New Testament Father.

Personal stories are rarely used; or if they are, they will be self-serving and demonstrate one-upsmanship The stories really do not have universal appeal or application. The Stork prefers linear story-telling and will not go off on tangents. S/he will refer to sermon notes extensively, perhaps reading an entirely pre-written text. Spontaneity is generally to be feared because things might get out of control or run longer than projected.

Probable Storks that I know:

____

____

____

____

Snowy Egret: Pastoral Version

Religiously, the Egret will choose a denomination that accentuates earning God's love and working toward perfection. An Egret will see God as a disciplinarian; s/he probably believes that one earns grace vs. grace being God's unfettered gift. The concept of forgiveness is difficult (except for oneself which the Egret has earned by being so watchful of everyone else). An Egret tends to be very concrete in his or her support of religious activities, focusing on details for an upcoming service vs. spiritual complexities. Current world events that are exacerbated by religious extremists are very hard for an Egret to process because s/he has difficulty grasping the whole continuum of believers within each faith tradition. On the other hand, if Easter events are in the hands of an Egret, every detail will be taken care of and the worship services will be flawless.

An Egret can be a religious zealot, braving all kinds of weather to be a Salvation Army bell ringer or going door-to-door to proselytize. S/he is holier than thou.

An Egret will always have his eye on the bottom line. When members of a church were asked to submit readings or inspirational thoughts for a prayer booklet, an Egret who must have been irritated by those who did not complete their pledges, contributed:

> *When thou vowest a vow unto God, defer not to pay it; for he hath no pleasure in fools; pay that which thou has vowed. Better it is that thou shouldest not vow, than that thou shouldest vow and not pay.*

EGRET RESUME STYLE

In order to help you sort through applicants for positions from whom you receive resumes, you may notice some of the following characteristics from an Egret. Facts will be precise, likewise format and spelling. Answers will tend to be sparse, short, often not giving the reader enough to flesh out the individual behind them. An Egret is not given to self-marketing, either because of modesty or lack of achievement. An Egret will avoid "I" statements. His/her handwriting is either the classic educational model of cursive or is very small, but readable. Choice of paper, where possible, will be beige and type fonts will not be attention-drawing. This is not a person who relishes graphic design, particularly not wanting to put his/her name in big, bold print.

If asked to comment on a personal weakness, the Egret may identify attention to detail as that area that is both a negative and a positive. Obviously, s/he will give it a positive spin, and probably will have struggled to do so.

EGRET SERMON STYLE

If you find yourself on a pastoral search committee (or any equivalent group selecting a new professor, college president, or non-profit high profile director), the following formal speech profile may be very useful.

The Egret's body language will be very stiff. There will be no lyrical motions, but s/he may use calculated punctuation or finger pointing. Any movements will be small, close to the body. S/he rarely will emerge from behind the podium. S/he will have a monotone, or a very carefully practiced modulation of tone for effect. Natural ease will not be demonstrated, either from the pulpit, in committee meetings, or casual conversations. An Egret carries him or herself with an erect body carriage and head tilted up. His or her nose mimics the Egret's pose. Most see this as an affectation and find the posture pretentious.

The Egret will follow the Lexicon for scripture readings and topics; his or her sermons will tend to be theological vs. real life dilemmas. S/he will cull stories from books or the Internet printed for the purpose. Rarely would an Egret tell of a personal experience, and jokes are inappropriate. What humor one sees is generally sarcastic, pointing out the foibles of others.

An Egret rarely will create his/her own prayers, choosing instead to use others' printed in reliable sources. S/he rarely is adept at children's sermons, as s/he lacks the interpersonal skills, playfulness, and use of stories or metaphors that make these seemingly simple sermonettes applicable for all ages. The language an Egret uses will be formal. His/her sermon will be well crafted, but it will lack personal warmth.

An Egret pastor may be an appropriate choice for a transition period that has been preceded by the emotional departure of his/her predecessor. S/he does not relish pastoral care duties or warm, interpersonal relationships with members of the church, but usually will deliver polished sermons (which s/he has been able to use in other short-term assignments).

Probable Snowy Egrets that I know:

____

____

____

____

Seagull:
Pastoral Version

The Seagull is really a floater. Often raised in a family that had ties to formal religion, the Seagull elects either not to connect at all, or to choose a faith group that is broader in its scope than some of the traditional Christian denominations. Bahai, the Unitarian Church, and the Reformed Jewish Temple will have appeal to a Seagull, as will the Episcopal Church, in which liturgy is ritualized and a hierarchy in the local church is clear and relatively undemanding of parishioners who choose to remain modestly connected. Again, the issue is affiliation — the active formal joining of a specific church.

SEAGULL RESUME STYLE

A Seagull would be an extremely rare pastoral candidate, though s/he might apply for an associate's position. In other business arenas, a Seagull typically will not have a resume ready to take advantage of an opening. S/he is not given to self-marketing or personal initiative for promotions, so s/he will have to scramble to create a resume. It will look much like an Egret's resume, without design or verbal flair.

Documentation of positions held will be precise. The Seagull's resume will be meticulous grammatically, but will not divulge many of his or her accomplishments. Because s/he is so used to seeing himself or herself as part of a dyad with a boss figure, it is very hard to take credit for independent achievement. Unless coached in the writing of his or her resume, the Seagull will be unduly modest or understated.

If a downsizing has occurred, s/he is an easy target because the company knows that the Seagull is unlikely to "make a stink" in ways that could benefit him or her and hurt the company. The old loyalty issue (falsely placed loyalty, in many cases) precludes a realistic appraisal of the layoff action. If you are in a position to hire a Seagull, read carefully between the lines and query pointedly to elicit the true accomplishments of this individual.

SEAGULL SERMON STYLE

It would be very rare for a Seagull to seek a position as a pastor, due to his or her distaste for being in the limelight and supervising other people. However, one does see Seagulls as lawyers, doctors, urban planners, and other professionals. The key is being able to be solitary in much of the work that s/he does. S/he is good one-on-one with customers, but does not like to do group presentations. The client relishes the very gentle and focused attention given by this able, organized professional.

Probable Seagulls that I know:

____

____

____

____

White Ibis: Pastoral Version

The White Ibis is frequently affiliated with a church or temple that focuses on local and/or international outreach. The key is that the Ibis is formally affiliated and can be counted on to serve in a variety of ways, whether that is through a local soup kitchen, parish care visitations, children's religious education, or provision of refreshments for social time following services. Tutoring disadvantaged students or immigrants and reading to children in homeless shelters are other ways in which White Ibises serve.

It is important to the Ibis that his or her whole family is involved. The Mormons and evangelical sects come to mind, as well as individuals in the mainline denominations. Many Jewish parents are highly involved in their temple's work in their community and ensure that their children progress through their bar mitzvah or bat mitzvah.

IBIS RESUME STYLE

As one reads through the resume of a White Ibis, often one will note prior work in one of the helping professions. For White Ibis ministers, there is a tendency to have trained and worked originally in social work, nursing, or educational fields before going on to seminary; but one will also see those whose careers in corporate or military fields left them searching for something of a spiritual nature.

A White Ibis resume and letter for a position will often contain very specific, individualized scenarios or anecdotes in response to the formal application questions. Words like "served," "provided," and "encouraged" will be used as the action words on the resume. S/he has a tendency to want credit for what s/he has achieved, but feels that it is immodest to take ownership for the completion of the task or project.

A White Ibis is not necessarily detail-oriented, so spelling may be problematic or sentences may be incomplete. Relationships and service matter more than form.

IBIS SERMON STYLE

A White Ibis will have a very warm, personal delivery with good eye contact and open, welcoming gestures and tone of voice. The stories used will often be childhood fairy tales, Aesop's Fables, and familiar biblical stories or personal stories from his or her own childhood. This approach can be very effective with children and some adults. But, men often have a subliminal negative reaction to a female White Ibis as if they are being treated like children. If this approach is used primarily with the children, then it is a safer technique.

The White Ibis has a tendency to scold and use "you" vs. "we." S/he sees God as a kind parent, but one who disciplines and expects adherence to the commandments. Grace is a harder construct for the White Ibis who sees love as conditional.

A White Ibis will move out into the congregation for christening and new member events. His or her language and sermon topics will be everyday experiences. S/he will follow the Lexicon readings, but will seek ways to make the message humble and less abstract.

Probable Ibises that I know:

____

____

____

____

Roseate Spoonbill: Pastoral Version

The Spoonbill will thrive best in a religious environment in which confession and forgiveness are pre-eminent. Confession allows the religious interface to be about the Spoonbill (i.e., personal attention for good or ill), and forgiveness allows him or her to "go and sin no more." Denominations that accentuate personal responsibility will not be as attractive to a Spoonbill.

Prayers of supplication are also important to a Spoonbill. Because s/he tends to live a life of extremes, peaks and pits, s/he constantly needs help from a power greater than himself or herself.

The drama of a liturgical mass, the scents of incense, and the almost magical nature of the high regalia of the clergy will appeal on many levels to the Spoonbill.

A parochial school environment that accentuates external discipline and rote learning will help willful Spoonbills concentrate on the lessons at hand and help them curb their tendencies to daydream and speak out before being called upon. They generally do not learn internal discipline.

SPOONBILL RESUME STYLE

In order to help you sort through applicants for positions from whom you receive resumes, you may notice some of the following characteristics from a Spoonbill. Most entries on a resume or in a letter of application will begin with "I." Answers to form questions will be free-flowing, going where s/he gets the greatest personal pleasure. Structure may be missing. Errata may be prevalent as s/he dislikes details. Misinformation may be provided, usually not with malicious intent, but rather due to sloppy record-keeping. Accuracy is not his or her "bag." However, there are Spoonbills who talk with classmates about how far they can safely stretch resume and bio information. This is intentional behavior.

Previous places of work may not show a normal progression of complexity and responsibility. Time at each workplace may be short with intermittent periods unexplained. The Spoonbill much prefers an oral history to a written biography or resume because his or her personal charm is a winning card.

SPOONBILL SERMON STYLE

The Spoonbill's sermon may have holes in its construction and delivery, just like his or her resume, because s/he does not value structure. S/he finds it confining and unspontaneous. S/he "knows" viscerally, artistically, and personally, but s/he does not seem to have much intellectual depth and breadth. Language will be colloquial as opposed to scholarly, and s/he will use conditional constructions frequently: "what might have been" or "what could have happened."

S/he may use stories of a personal nature; while that can be a very instructive and personally disclosing technique in the hands of some other birds, it often fails for a Spoonbill because listeners fail to see the relevance to the morning's lesson. Because a Spoonbill cherishes spontaneity, s/he may not even use cue cards, with disastrous results when s/he forgets the ritual words in prayers or in the Scripture that has just been read. Sermon coaches have their hands full with Spoonbill preachers who lack the personal cognitive discipline to do without practice and notes.

The Spoonbill may use humor inappropriately in a sermon, having been amused by something s/he has found on the Internet which is only tangentially connected to the topic. Also, s/he has a tendency to forget the punch line!

A Spoonbill will favor creative approaches to worship, including liturgical drama, art, video, music, dance, etc. As such, s/he adds variety and opportunities for those who learn in non-lecture formats. However, these approaches will often suffer from his or her free form tendencies, so there will not be as much cohesion as a Great Blue Heron would bring to the same approach.

Gestures and vocal tones may seem uncontrolled or out of sync with the topic. Often they are very dramatic. Conversely, when a Spoonbill is at a low point emotionally, his voice will be monotone and gestures lifeless. His or her choice of ministerial "garb" is artistic. His or her appearance may not be mainstream, with a beard or mustache for men and a hairstyle or color for women that is expressive, not traditional.

Pastoral visitation will probably not be a strong point with a Spoonbill. Not used to focusing on others, and being rather inept at conversations of spiritual depth, the Spoonbill will feel like a fish out of water and will make his or her patient feel very uneasy.

Probable Spoonbills that I know:

____

____

____

____

Brown Pelican: Pastoral Version

The Pelican is rarely found in the ministry. There is an inherent discipline that typically would not be attractive, except perhaps to a Fair Pelican.

In terms of affiliation, a Pelican would enjoy a setting that allows significant social interaction. When religious groups feature outreach and projects such as bazaars and men's fellowship breakfasts, the Pelican will enjoy being a player (but probably should not be a key planner). S/he will insert fun into the event, which goes a long way in helping others get through the drudgery of some of these projects.

PELICAN RESUME STYLE

In order to help you sort through applicants for positions from whom you receive resumes, you may notice some of the following characteristics from a Pelican. The format will be attractive, as the Pelican values an aesthetic presentation. The pronoun "I" may be prevalent, as s/he generally is immodest. His or her responses to form questions may be very creative, as may be the grammar and spelling. The employment history may show a number of short-term jobs. When queried about this pattern, s/he may note that s/he had some wonderful opportunities to learn in these various settings, so leapt at the chances to explore new skills, techniques, or approaches.

There may also have been some radical shifts in career fields, from insurance sales to the ministry, for example. The Pelican is a maverick. His or her commitment is to get the most possible out of life; s/he often does not demonstrate commitment to a specific company or mission, much to the annoyance of some of the other species.

PELICAN SERMON STYLE

The Pelican may be a delightful change in pace from previous ministers, but s/he generally is not a candidate for a long term position (both because s/he relishes change and because the congregation may yearn for greater depth after awhile). The Pelican will almost always have a smile on his or her face and s/he will use humor frequently in sermons, regardless of its appropriateness. Often his or her stories or jokes have come from the comic strips or movies. His or her breadth and depth of thought can be narrow and shallow. S/he may have a fresh approach to familiar topics, but there may not be a consistency of viewpoint over time.

Gestures will be ebullient. Eye contact will be friendly. Tone of voice and facial expressions will be varied. His or her vocabulary is enormous, but malapropisms do pop up from time to time. S/he is a great story-teller and has particular appeal for children. While other species must work very hard to craft children's sermons that are playful and instructive, the Pelican is a natural. S/he will include himself or herself in the lesson, not setting himself apart from the congregation.

If s/he gets on a roll, s/he may deviate from his or her loosely designed sermon, and may risk ending with a whimper instead of a bang. Ending on schedule will not be important to a Pelican preacher.

Probable Pelicans that I know:

Green Heron: Pastoral Version

For a Green Heron, religion is generally distrusted, unless it has become one's area of study. For the lay person, a denomination that operates with ritual and tradition, such as the Episcopal church, may provide sufficient assurance that s/he will not be challenged emotionally. There is rarely the do-gooder fervor for social justice visible in weekly services, so the Green Heron can feel safe in the standard liturgy. Other Green Herons, like Ralph Waldo Emerson, are attracted to the cerebral Unitarian church.

A Green Heron pastor can do home and hospital visits, but they will be short. S/he is extremely uncomfortable, sensitive, and stiff in these situations. S/he generally does not initiate conversation, but can withstand awkward silences. Others may perceive him or her as aloof.

Committee work in a church setting is not a favorite venue for a Green Heron. S/he much prefers to listen to others, then work solo and share results with a committee for final approval.

Parishioners who want an affable, easy-to-know pastor who is very approachable and supportive on a personal level may find the Green Heron stand-offish. Those who spend time getting to know him/her will find a deeply sensitive, committed soul.

Some of our greatest theologians have been those who moved beyond parish life to a seminary, monastery, or the equivalent, where intrusions of the daily have been minimized and the focus is on the spiritual, the immortal.

GREEN HERON RESUME STYLE

In order to help you sort through applicants for positions from whom you receive resumes, you may notice some of the following characteristics from a Green Heron. His or her language will be scholarly, literary, and accurate. S/he is not comfortable marketing himself or herself. She will avoid the pronoun "I," but probably will give statistics on achievements (i.e., s/he will quantify the results of his or her work). The resume may feel like an encyclopedia account.

The Green Heron is easily confused with the Snowy Egret in a resume, because the answers or comments may be short, cryptic, and modest. However, one telling difference is the intellectual focus of the Green Heron. S/he may digress from a form question which s/he perceives as lacking in depth or breadth, so read the answer carefully. Wit may be displayed if triggered by form questions.

GREEN HERON SERMON STYLE

A Green Heron generally dislikes public speaking enormously. Therefore, when s/he finds himself or herself in a profession that requires weekly sermons, s/he structures a sermon analytically. It is rarely colored by analogies, metaphors, parables, or anecdotes. It will be cohesive, dry, and short. References will be scholarly. The entire sermon is often written out vs. simply outlining key points to be made. There can be a tendency to read the sermon, because s/he has worked so hard to get just the right wording. When s/he tackles complex issues of the day, such as social justice, sexual preferences, sexual abuse by the clergy, discrimination, etc., the approach will be clinical, not personal. For parishioners who want larger issues translated into actions that they may take, the Green Heron believes his or her role is to provide the theoretical exegesis (the forest); it is up to the listeners to see the trees and determine desirable actions.

The Green Heron is generally not an adept impromptu speaker (or giver of prayers), though s/he may grow into this with practice. His or her own prayer life will be intense. His or her public prayers will be keyed to the sermon theme, as will the hymns. Sometimes the hymn choices are rather obscure works because the text fits so well.

The Green Heron minister will tend to be polishing his or her sermon until the very last minute. Again, technology has enabled this ongoing tweaking of its contents. It is not necessarily the result of failing to work on it soon enough; rather, it is a compulsion to make it as good as possible and his or her brain continues to tinker with the concepts and comments until the time of delivery. Where other birds reach a "good enough" level, the Green Heron demands ever more of himself or herself.

The Green Heron would prefer to remain in the pulpit vs. moving around the congregation during the service. S/he does not do much with arm gestures. His or her body is fairly stationary and undemonstrative. His or her vocal tone is fairly constant, with not much variety. S/he does not emphasize or punctuate much vocally, but s/he may do so via the grammatical construction.

Usually, a Green Heron is not at ease with children's sermons because s/he lacks the warmth, playfulness, and ability to put his or her communication into a style or plane that children relish.

However, the negatives mentioned above are often ameliorated when the preacher is a Fair Fowl. S/he has learned to extend herself or himself into the interpersonal domains and has made herself/himself more accessible and approachable for individuals and the congregation.

Probable Green Herons that I know:

____

____

____

____

Great Blue Heron: Pastoral Version

The Great Blue Heron will have a mission or calling that is beyond his own life. That may be a quest such as Don Quixote's or it may be that of Jesus Christ. There will be a mixture of championing the poor and the weak, righteous indignation with the smug, greedy, and hypocritical (money lenders in the temple), and courageous, bold anger. There will be a mix of divine discontent and divine discomfort.

The Great Blue Heron may be affiliated with a formal church, or s/he may stand outside of denominational teachings on a more philosophical plane. But, his or her stance will be very much like Erma Bombeck's stance: "When I stand before God at the end of my life, I would hope that I would not have a single bit of talent left and could say: 'I used everything you gave me.'"

George Bernard Shaw expressed a similar sentiment:

> *This is the true joy in life, the being used for a purpose recognized by yourself as a mighty one; the being a force of nature instead of a feverish, selfish little clod of ailments and grievances complaining that the world will not devote itself to making you happy.*
>
> *I am of the opinion that my life belongs to the whole community, and as long as I live, it is my privilege to do for it whatever I can.*
>
> *I want to be thoroughly used up when I die. For the harder I work, the more I live. I rejoice in life for its sake. Life is no 'brief candle' to me. It is a sort of splendid torch which I have got hold of for the moment — and I want to make it burn as brightly as possible before handing it on to future generations.*
>
> George Bernard Shaw,
> *Man and Superman*

A Great Blue Heron cleric must be careful on two ends of his or her continuum. She must not be perceived as a White Ibis because of her warm, personal, motherly stories. This type-casting can diminish her chances to move into an appropriate leadership position. On the other hand, a Great Blue Heron with enormous creative discipline and energy may be perceived as too much of a showman. This multi-gifted individual can get carried away with his own enthusiasms. However, s/he does forever set the level of expectations at a new norm — hence, there is great difficulty in finding an appropriate successor.

BLUE HERON RESUME STYLE

S/he is comfortable with self-marketing, so sh/e will ensure that his or her materials look good, are edited, and are accurate. Details do matter to the Great Blue Heron, both because of self-integrity and a belief that others deserve the best version deliverable on time. (This detail focus is different from procrastinating due to detail- or micro-managing.) There will be a display of style, even within the constraints of some pre-ordained resume and application formats.

S/he will tend to include in an application packet materials that flesh out his or her responses. For example, publications, videos/ CDs of performances, and newspaper articles about his or her work will be sent, so that others are commending his or her work or the recipient can judge for himself. In an attempt to avoid the pronoun "I," s/he will use the third person to describe activities and achievements (i.e., she or he).

There will be an accurate description of his or her role in accomplishments because s/he does not believe in aggrandizing or in undue modesty.

BLUE HERON SERMON STYLE

The Great Blue Heron will use global language with a Latinate literary quality vs. the basic Anglo-Saxon words in our language. S/he likes the subtle nuances of language (vs. a scholarly approach). This is the pastor who will regale you with the ever-so-slight, but meaningful differences in the words that describe "joy."

The Great Blue Heron will have a graceful style of writing and speaking, often rather poetic or musical. Cadences will be used for repetitive power (such as those used by Sojourner Truth and copied later by Martin Luther King and recent politicians). Alliterative structure will be used, either in sentences or in concepts. Words beginning with the same sound are easier for people to remember and demonstrate a careful creative construction.

S/he likes and uses parables and metaphors, as have a long line of literary and religious leaders, from Buddha and Jesus and Aesop (of fable fame) to modern day writers about cheese being moved! A *USA TODAY* article on the "Cheese" books notes the polarity between those readers who "get it" and those who avoid such writing altogether. If the congregation is made up of very concrete thinkers, the Heron may need to moderate this technique.

S/he uses stories, often personal, to serve as parables and uses harmless humor to great advantage. S/he values a variety of ways to share God's message, so sh/e may use liturgical drama, dance, art, and music to enhance worship. S/he is widely and deeply read.

The Great Blue Heron's use of the word "we" is inclusive. S/he includes herself in the lessons being taught and will refer to "our" journey of faith.

Vocal tones are musical. His or her pulpit voice is an extension of his regular, natural voice, but is slower to ease hearing and understanding.

Gestures are appropriate — often with arms outspread and open. S/he will move away from the podium, allowing the congregation to see the whole person. When s/he moves into the congregation, it is in a communal spirit (for welcoming new members or baptising a child).

S/he likes to match the children's sermon to the pulpit message, understanding the cumulative power of a message delivered by several learning stages and styles. S/he is at ease with the children and they reciprocate, with often highly insightful responses to his or her queries.

The Great Blue Heron would see the Lord's Prayer as a request, a plea in the subjunctive mode (vs. a demand). S/he sees God as a father (parent) "who gave His only begotten son." Grace is a concept with which s/he is conversant and convinced.

S/he will not necessarily follow the Lexicon for Scripture readings and topics, if his or her sense of situational leadership leads to other more pertinent, timely topics.

Probable Blue Herons that I know:

____

____

____

____

Scenarios

Scenarios

These scenarios will provide a quick sketch of each bird's *prototypical response.*

APPROVAL VS. RESPECT

Stork	Expects respect; gives approval (i.e., permission) to others once they have proved actions are justified.
Snowy Egret	Deigns to give approval, but very reluctantly as it lessens his or her control to a degree. Approval will come with unsolicited critique and advice. Seeks approval from supervisor when uneasy about outcome.
Seagull	Wants respect from key others and approval of priorities for his or her work.
White Ibis	Expects respect, because sees self as selfless. Gives conditional respect, if other does as required, but will point out how the effort falls short and fix it.
Spoonbill	Expects demonstrations of respect from others through receiving the best of everything in the workplace. Does not seek approval prior to actions.
Pelican	Values pleasurable experiences, of which respect is one. Seeking approval would limit spontaneity.
Green Heron	Wants respect for his or her ideas; does not like seeking approval for projects.
Cormorant	Fails to understand the concept of respect for self or others. Purposely avoids seeking approval.
Blue Heron	Values respect for self and others. Has learned that approval comes with strings attached, for good or ill.

ARGUMENTS AND ANGER IN THE WORKPLACE

Stork	Loves arguments. Is provoking, intransigent, intimidating. Must win. When angry, is given to outbursts and yelling. Threat of physical violence is just below the surface.
Snowy Egret	Snipes from the side lines. Stabs with negative criticism. Demands that you remain until work is satisfactory, regardless of time.
Seagull	Avoids arguments at all costs. Frustration will be demonstrated by passive-aggression. You won't be able to obtain the disk, key, or whatever it is that you need to complete a task according to your desires.
White Ibis	Takes martyr stance. Stabs at others below the belt, all the while saying s/he wants harmony. Will cry and then provoke.
Spoonbill	Is provocative and daring in comments. When trapped, will lie and cry.
Pelican	Is a freethinker and is often oblivious to his or her impact on others. When angry, will lash out at source of hurt.
Green Heron	Is thought-provoking; offers critical thinking negatively. Avoids public arguments whenever possible. When angry, is unavailable in person or in emotion.
Cormorant	Is just plain provoking; seeks others' humiliation. Once s/he gets ahold of someone, won't let go, because s/he must win at all costs. Must have an adversary to survive.
Blue Heron	Is thought-provoking in a diplomatic way; seeks others' elucidation, but is capable of changing his or her point of view based on others' input. Likes negotiation. Dislikes confrontation and will avoid, if possible. If confronted, will get very quiet and will stab verbally when strategic.

ATTRIBUTES

Stork	Commanding, spontaneous
Snowy Egret	Diligent, precise, prim, disciplined
Seagull	Loyal, efficient
White Ibis	Nurturing, manipulating, martyr-prone
Spoonbill	Melodramatic
Pelican	Hedonistic, pleasure-seeking
Green Heron	Cognitive-focused
Cormorant	Devious, flaw-focused
Blue Heron	Goal oriented, effective

AVOID THESE FOLKS WHEN:

Stork	S/he lost control, was upstaged, or was reprimanded today.
Snowy Egret	S/he has not done his or her personal best today (so is angry with self and lashes out at others who are also imperfect.
Seagull	S/he was taken for granted today or his/her boss was mistreated/threatened today.
White Ibis	S/he has not been appreciated for his/her tireless efforts on others' behalf today, so is in a martyr mode.
Spoonbill	S/he didn't feel "special" today.
Pelican	S/he has not had time for fun today. All work and no play makes a Pelican sad today.
Green Heron	S/he was interrupted, and his/her ideas, concepts, or data were not valued today.
Cormorant	S/he didn't "one-up" someone today.
Blue Heron	S/he was not recognized for his/her vision and respect for others today.

AWARDS AND EVALUATIONS

Stork	Likes the bell curve (or less!); has a hard time believing that workplace can have many exceptional people. Will be very public in giving recognition.
Snowy Egret	Gives only to those whose work is perfect, but wants praise personally. Sets very precise and hard-to-achieve standards and holds employees to them. No "favoritism" or warm praise; emphasis is on areas in which to improve; stingy with praise and awards.
Seagull	Wants personal praise and recognition, including monetary. Can be stingy with praise for others.
White Ibis	Gives awards conditionally as a mode of control. Gives brief explanation or assessment; is not generous with praise. Rarely initiates on-the-spot awards unless a senior suggests.
Spoonbill	Wants rewards regardless of effort or product. Wants to be sure others are not getting more. Does not initiate awards for others. Sees a heavy emotional component to evaluation (i.e., how much support s/he perceives from employee; what have you done for me vs. company? lately?)

Pelican	Wants frequent praise. Is innovative in giving rewards (such as a fancy dinner for whole work team). Wants painless process for supervisor and those supervised.
Green Heron	Wants private, not public recognition, and only from those whose opinion counts. Is usually not good about praising others. As supervisor, will be fair and praise in a muted, official way. Likes analytical rating system; observes employees so quietly that they often wonder where they stand with him or her.
Cormorant	Sees award as a competition to be won. Therefore, is not likely to award others unless there is some benefit to be accrued personally. Is not regular about supervisory duties.
Blue Heron	Praises whatever and whenever folks do well. Believes many in workplace meet level of excellence in their work. As supervisor, sets more global levels of achievement and assesses that way as well. Accentuates the positive. Holds each employee to his or her own goals. Doesn't see a need to compare to others... "personal best."

BODY LANGUAGE "ON THE BEACH" WHEN APPROACHING OTHERS

Stork

S/he walks right toward other birds and doesn't slow his or her pace. S/he expects them to move out of his/her way. Or, s/he runs past the birds, then stops suddenly to look at them, which the other birds perceive as threatening.

Snowy Egret

S/he seeks to identify the species and count the number. S/he assesses if any Egrets are attempting to take part of her territory, in which case s/he becomes very feisty and vocal, head feathers all fluffed up.

Seagull

S/he is generally disinterested;
s/he has his/her own focus, which is different from the rest.

White Ibis

S/he walks around the others, gently trying not to disturb.

Spoonbill

S/he worries about others that might distress him/her.

Pelican

S/he laughs at the antics of the other birds, finding amusement in their behavior.

Green

S/he observes the other birds without moving.

Heron

Binoculars and camera are favorite tools of the human variety.

Cormorant

S/he maliciously chases other birds.

Blue Heron

S/he stands at a distance and observes, then walks further away, preferring to be in deeper water or solitary.

CHANGE

Stork	Likes change as long as s/he initiates and controls.
Snowy	Dislikes change because it disrupts his or her Egret labored schedules, rules, protocols, policies, etc.
Seagull	Once s/he's been able to research and organize the change, s/he's in control; but s/he is not good with spontaneity.
White Ibis	Accommodates if change is good for those dependent on him or her.
Spoonbill	Worries about the implications for him or herself; therefore, is slow to change.
Pelican	Loves change; is buoyant. Loves adventure and implementation.
Green Heron	Wants to be part of decision-making to ensure change is fruitful.
Cormorant	Thrives on chaos; brings morsels of opportunity. Also, change helps him or her not to have responsibility for completion of previous tasks.
Blue Heron	Sees opportunity in change, but knows that pace mustn't be too frenetic to prevent vision and strategic thinking.

CHARACTERISTIC PHRASES

Stork	"Just do it my way!" "I want..........."
Snowy Egret	"Prove it!" "Show it to me in writing."
Seagull	"I read your mind, didn't I?"
White Ibis	"Let me help."
Spoonbill	"I feel... and I need..."
Pelican	"What fun!"
Green Heron	"I think......"
Cormorant	"I disagree......"
Blue Heron	"How can we use/do it?"

Control and Satisfaction

Stork	Seeks control over others.
Snowy Egret	Seeks control over self (and others, if s/he can manage it).
Seagull	Seeks to under-control self in order to combine with others.
White Ibis	Seeks vicarious satisfaction through the achievement of others whom s/he controls/ supports.
Spoonbill	Seeks uncontrolled self-pleasing satisfaction.
Pelican	Seeks satisfaction via over-active stimulus/ achievement.
Green Heron	Seeks satisfaction via individual thoughts.
Cormorant	Seeks a combination of satisfaction and dissatisfaction with self and others; references other authority figures.
Blue Heron	Seeks satisfaction via his/her own and others' achievements (those whom s/he has supported).

CUMULATIVE WORKPLACE STRESS

Stork	Gives orders and more orders; tries to bully others into action. Will attack others frivolously, because any way s/he can gain control feels better than feeling powerless.
Snowy Egret	Tries to maintain perfection in all arenas of activity, regardless of their value. Protects his/her turf.
Seagull	Hunkers down and initiates even less than usual; is physically present, but emotionally absent.
White Ibis	Becomes a martyr. Makes sure that those s/he has supported know of her hurt and malaise.
Spoonbill	Gets very pale (vs. pink) and very dysfunctional. S/he seeks attention, for good or ill.
Pelican	May move on if there is no fun. If s/he can't leave, may become depressed.
Green Heron	Hunkers down and focuses on analytical domain only. Closes others out.
Cormorant	When pool becomes shallow, his/her activities become more visible and s/he becomes more reckless.
Blue Heron	Tends to burn out attempting to buffer others' impact and attempting to provide enlightened leadership when others don't have the energies for reciprocal followership.

DECISION-MAKING

Stork	Makes a decision — usually promptly. Can be a shoot from the hip move. Has a hard time admitting a mistake and changing course.
Snowy Egret	Can take forever — has to be perfect.
Seagull	Decisions for work can be prompt, but decisions for self often very slow.
White Ibis	Plans a long time in advance and gets a great deal of pleasure from the process.
Spoonbill	Hates the process; wants a decision.
Pelican	Is drawn to variety and all-you-can-eat smorgasbords.
Green Heron	Usually studies/shops for most elegant solution; can take so long that s/he is overcome by events.
Cormorant	Frenetic decision-maker; shoots from the hip, then changes his or her mind/course frequently and can't remember why s/he made the original choice.
Blue Heron	Prefers decision and action, but can wait a long time, if strategic. Believes in doing best s/he can do for now, and then regrouping, if necessary.

EATING HABITS (RESOURCE CONSUMPTION)

Stork	Likes a big helping of expensive foods to show status.
Snowy Egret	Likes small portions of normal fare; likes to control the amounts served to others.
Seagull	Doesn't prepare food for others often; is willing to eat whatever is easiest, except for truly special occasions.
White Ibis	Controls the amounts of food for others; food is his/her milieu and s/he looks for appreciation in this domain.
Spoonbill	Likes lots of sauces, pretty colors, and chocolate.
Pelican	Is drawn to variety and all-you-can-eat smorgasbords.
Green Heron	Is very controlled in the amount s/he consumes; is willing to help prepare the food, especially if s/he can use tools.
Cormorant	Takes more than his/her fair share.
Blue Heron	Prefers to let guests/family self-serve, taking what they want of attractively presented fare. Enjoys variety and a colorful mix.

EMPLOYEES NEED

Stork	Stringent boundaries.
Snowy Egret	Opportunity and time to get things in order.
Seagull	Support, especially emotionally.
White Ibis	Opportunity to mentor or care for others.
Spoonbill	Opportunity to feel special.
Pelican	Variety of tasks/opportunities.
Green Heron	Opportunity and time to think.
Cormorant	Opportunity to feel secure, non-threatened.
Blue Heron	Opportunity to be visionary and be praised.

EXTERNAL THREAT TO ORGANIZATION

Stork	Will fight tooth and nail. Sees as a win-lose situation. Not so concerned with the impact on people as with impact on power base, reputation, status.
Snowy Egret	Says, "Show it to me in writing. They have to follow rules."
Seagull	Is concerned with impact on leader(s) whom s/he serves.
White Ibis	Takes emotional care of others in organization.
Spoonbill	Calamitizes. Oh, woe is me. It will be awful. We'll lose everything.
Pelican	May find very creative solution. Won't worry about the outcome. "Qué sera, sera." Will deal with it when it comes; meanwhile, no investment of energy or emotion in the situation. Will seem cavalier, blase to others.
Green Heron	Believes loopholes may be found to preclude threat if given time to study situation.
Cormorant	Sees opportunity for self in chaos. Not concerned about others.
Blue Heron	Looks for strategies for a win-win. Is concerned about the impact on employees and on the mission of the organization.

FRIENDSHIPS

Stork	Camaraderie of titans; doesn't quite trust the others, so is always a little wary, but this is the classic "old boys' network." Likes to do action/ competitive sports with friends.
Snowy Egret	Has few friends; asks a high level of loyalty and perfection. Few measure up.
Seagull	Often underestimates how many others value him/her. Tends to focus on a few very long-time friends, though enjoys lunch-time sociability. Doesn't tend to join churches, clubs, etc.
White Ibis	Has many acquaintances and a numerous life-long friends upon whom s/he can count. Has warm, generous relationships. Can be inclined to let others take advantage of his or her generous spirit. Also, can be a little suffocating for more reclusive types.
Spoonbill	Has a number of devoted close friends. Especially enjoys White Ibises.
Pelican	Lots of acquaintances and a number of long-term friendships that s/he puts more into than may be reciprocated. Is deeply hurt when others are not as responsive or thoughtful as s/he is.

Green Heron	Has few acquaintances and friends (and they need to initiate most of the interactions in the relationship). This very reclusive person is most comfortable in his/ her own company.
Cormorant	Knows a lot of people and is outwardly sociable, even playful. But, doubts that they would ever really support him/her in a pinch because s/he knows s/he wouldn't unless personally advantageous. Wants loyalty, but doesn't give it. Always suspects motivation of others.
Blue Heron	Has many acquaintances and a few life-long friends, often of both genders. Has platonic friendships as a young person. Enjoys being with people, but also needs alone time. Diverse acquaintances, often includes those mentored. Puts more into relationships than most can reciprocate.

HOUSEKEEPING IN THE WORKPLACE

Stork	Spaces are usually immaculate, but s/he tries to get others to do it for him or her. If s/he knows that superiors will inspect, s/he is a tiger for cleanliness. But, it's more a control-of-others focus — wants a clean desk because s/he has delegated everything.
Snowy Egret	May tend to be spartan. Will ensure his or her own space is meticulous. Will also brush lint/dandruff off of others. Always has a lint roller in the drawer. Will use time cleaning to avoid tasks whose outcomes are not as clear. Desk will be neat, if not perfectly clear, at the end of each day.
Seagull	Literally picks up after all of those who don't clean up their own spaces. Wants praise as a result.
White Ibis	Spaces will be clean, but s/he may have a lot of belongings and piles of work to be done. S/he feels put upon to do more than others.
Spoonbill	Usually takes a great pride in artistic appearance of his or her spaces; they announce the values of the individual who works there. S/he resents if others dump stuff in his or her area in order to get their own areas neat under pressure from "on high."

Pelican	Generally is oblivious to the state of his or her spaces, because focus is on fun, not cleaning. The way to involve a Pelican is to make workplace cleaning a group project that features fun.
Green Heron	Can be exceedingly neat, if space is public; but, can be an exceptional pack rat. Neatness is not valued per se, but being able to find something s/he needs is. Will not willingly join in general workplace cleanup; stays out of sight. Believes s/he has more important things to do.
Cormorant	Often soils his or her own spaces. May sometimes use space as entrapment for others (with scents, low lighting, and music), but sometimes these accoutrements are for his or her own cocoon to calm constant anxieties.
Blue Heron	Wants space to look good — both clean, orderly, and attractive. May have lots of projects in progress, but can find what s/he needs. Helps others clean spaces (though wants to be praised for doing so), and picks up litter/messes wherever s/he sees them, regardless of cleaning schedule or responsibility. Is concerned with workplace image.

INFORMATION SHARING VS. WITHHOLDING

Stork	Believes that information is power; therefore withholds on principle. However, in situations when s/he feels challenged or sees opportunity to gain status, will share portions of the information, erroneously believing that s/he has maintained its secrecy.
Snowy Egret	Withholds information as a given. Then s/he questions its value and insists that it be documented.
Seagull	Withholds information because s/he is not sure what portions may be critical to his or her superiors. Sometimes hints that s/he knows more than s/he is sharing.
White Ibis	Shares information conditionally. Would expect others to reciprocate.
Spoonbill	Seeks what is in any information for him or her. Shares if it will bring personal rewards.
Pelican	Feeds on information and is willing to share with all without constraint.
Green Heron	Withholds information until it is part of a total concept. Does not share daily information that comes his or her way, mostly because s/he is not interested in it.
Cormorant	Withholds information to parlay and manipulate as best suits him or her. S/he recognizes that some of the profiles feel "pressed to talk" and gain status, and will, if given enough latitude, give away key morsels that can be pieced together. S/he knows that pride goeth before a fall, or at least, a stumble.
Blue Heron	Sees information as a commodity to be shared as part of strategic planning.

LANGUAGE AND TONE

Stork	Magisterial tone; commanding language.
Snowy Egret	Sniping, critical tone; precise language.
Seagull	Emotion-free tone; efficient use of words, as few as possible.
White Ibis	Emotion-laden tone; use of conditional mode ("should" and "ought").
Spoonbill	Whining tone; also uses "ought" in terms of expectations from others.
Pelican	Excited tone, with expectations of delight; loves power of unusual words and puns.
Green Heron	Reserved tone; technical language in terse phrases.
Cormorant	Manipulative tone; usually talks more than necessary and directs all attention to self.
Blue Heron	Speaks expressively, directly, and chooses words carefully, concerned about their impact on listeners; has a musical, rhythmic cadence and enjoys alliteration.

PERSPECTIVES

Stork	Win - lose (s/he must win)
Snowy Egret	Win - lose (s/he wins most of the time through sheer determination and referencing of the rules)
Seagull	Lose - win (s/he is willing to lose in order to keep harmony and favor with his/her superior, but will insist on winning to protect boss)
White Ibis	Win - lose (s/he must win through control of others and provision of final touches to bring up to his/her standard)
Spoonbill	Lose - win (s/he believes s/he loses most of the time, that others get better treatment/ stuff/opportunities)
Pelican	Win - win (there are so many options to pursue and so much talent to be successful)
Green Heron	Win - lose (his/her ideas must be accepted en toto to win)
Cormorant	Lose - lose (and s/he always gets the short end of the stick, from his viewpoint, so s/he works to damage the winner, even though it muddies him/her in the process)
Blue Heron	Win - win (working for common goal permits many to be their own personal best)

POWER MOTIVATION

Stork	Pragmatic power
Snowy Egret	Power of perfection
Seagull	Protective power
White Ibis	Parental power
Spoonbill	Personal power
Pelican	Power of pleasure
Green Heron	Philosophical and intellectual power
Cormorant	Perverted power
Blue Heron	Persuasive power and power of praise

PRIDE

Stork	Arrogance
Snowy Egret	Ethical superiority
Seagull	Reflected glory
White Ibis	Vicarious pleasure through others' success
Spoonbill	Prowess in sensitivity
Pelican	Pleasure in products
Green Heron	Cognitive delight
Cormorant	One-upsmanship
Blue Heron	Pleasure in process with positive impact on people; delight

PRIMER

Stork	Decider
Snowy Egret	Derider
Seagull	Survivor; Consigner
White Ibis	Provider; Co-signer
Spoonbill	Whiner; Sigher
Pelican	Diner
Green Heron	Miner; Archiver
Cormorant	Conniver; Maligner
Blue Heron	Diviner; Designer

QUESTS

Stork	Conquestor (seeks own status and power)
Snowy Egret	Inquestor (seeks to find something wrong)
Seagull	No-questor (receives vs.seeks; scavenges from available resources)
White Ibis	Bequestor (controls by giving)
Spoonbill	E-questor (seeks emotional component); also a Protestor
Pelican	Me-questor (hedonist)
Green Heron	Sequestor (seeks privacy of mind and space)
Cormorant	Detestor (resents others' success and festers)
Blue Heron	Respector, Key-questor, and We-questor (seeks the proverbial Holy Grail and values those who travel with him or her)

RULES AND REGULATIONS

Stork	Rules are for other folks. His or her power is above the peasants. Do now and ask forgiveness later (if necessary). Abides by the law when it is convenient and advantageous.
Snowy Egret	Workplace manual or Bible contains concrete rules to befeared and followed. Do not question, just follow to the letter. Society needs this absolute structure.
Seagull	Knows the rules and protocols, and follows when they meet the needs of his or her senior.
White Ibis	Follows his or her own rules which are usually delivered in a quiet but manipulative way. S/he requires loyalty from those s/he supports. Sees the law as cold and impersonal.
Spoonbill	"I need; I want" are the only rules. "Fairness" is getting a little more than others. Is generally oblivious to laws, rules, and regulations.
Pelican	Dislikes constraints. Wants freedom to experience and decide.
Green Heron	Abstracts or process rules; needs freedom to analyze.
Cormorant	Rules are for the foolish; but s/he tries not to be visible in flaunting them.
Blue Heron	Rules are starting points for negotiation, not absolutes. Respects the law, but will work to change it if s/he sees as no longer valid.

STRATEGIC RESPONSES TO:

Stork	"Yes, Sir!"
Snowy Egret	"You're right........"
Seagull	"I couldn't get along without you. Thanks for holding everything together."
White Ibis	"Thank you for your help and concern. I really appreciate it."
Spoonbill	"Your work is very expressive."
Pelican	"Your multi-talents amaze me!"
Green Heron	"Your interpretation and summary are on the mark — concise."
Cormorant	"I like your position on..." (use this comment to confound and confuse him or her when s/he has intended a comment to wound).
Blue Heron	"What an impressive idea/project/vision!"

STRATEGIC SEATING IN RELATION TO THESE TYPES

Stork	Sit to his left, so s/he is forced to use right brain as s/he speaks with you. This will help diffuse some of his or her shoot-from-the-hip responses. S/he would prefer to confront you face-to-face.
Snowy Egret	Sit to his left, to diminish the Egret's left brain activity and activate a broader perspective for his or her efforts. S/he would prefer to confront you face-to-face.
Seagull	Sit across from the Seagull because s/he likes your undivided attention. S/he wants to read your body language so that s/he can anticipate your needs (both left-brain sequential needs and right-brain emotional needs).
White Ibis	Sit to his left, so that s/he can respond from his or her preferred right brain. You don't anticipate being confrontational with this person and will accomplish more through comfortable interaction. You will be using your left brain, which is a good counter-balance.
Spoonbill	Sit to his/her right, so that s/he must respond with his/her less dominant side of the brain, the logical/sequential. This will help the Spoonbill to be more structured. S/he would like you to sit on his/her left, to use her right brain.

Pelican	Sit across from a Pelican, allowing him and you to access both sides of your brains. This enhances the integration of the spontaneous with the structured. S/he would like you on his/her left, to use his right brain.
Green Heron	Sit to his left, because s/he will be forced to incorporate some right brain thinking into his/her left brain analysis. S/he would prefer you on his right.
Cormorant	Sit to his right, because this will trigger his left brain analytical processes which his emotions may have overpowered.
Blue Heron	Sit across from him, allowing you both to utilize both sides of your brains for the most effective, imaginative solution. S/he enjoys your eye contact.

SUCCESS BY SOMEONE ELSE IN THE WORKPLACE

Stork	Damn! S/he beat me/us to it. What will it take/cost to catch up?
Snowy Egret	If I wasn't involved and key to the result, then I am (a) mad; and (b) I will find a way that s/he abrogated the rules.
Seagull	How does that help us? Is there a benefit for the whole group?
White Ibis	Success for our group = success for all of us.
Spoonbill	What did I get out of it?
Pelican	Hey! Well done! What can I learn from it?
Green Heron	Let's see if it works? lasts?
Cormorant	Oh, damn! S/he got all of the glory!
Blue Heron	Yeah! Wow! Good show! Good for the cause! Go, Buddy!

TIME ORIENTATION

Stork	Past
Snowy Egret	Past
Seagull	Past
White Ibis	Present
Spoonbill	Present
Pelican	Present
Green Heron	Infinite Time of ideas
Cormorant	Present
Blue Heron	Future

WORKPLACE BUDGET

Stork	Money is power; therefore, holds tightly and shares only partial information.
Snowy Egret	Manages what is allotted meticulously.
Seagull	Manages with accuracy, but enables others to get what they need.
White Ibis	Doles out small portions with strings attached and expects gratitude.
Spoonbill	Is carefree with money. Spends and then expects more than others.
Pelican	Money is a passport to pleasure.
Green Heron	Is fiscally conservative and secretive.
Cormorant	Focuses on getting the biggest share of the money.
Blue Heron	Shares information with subordinates and requires that they build their own budget segments. Has the vision to request an amount needed for innovation.

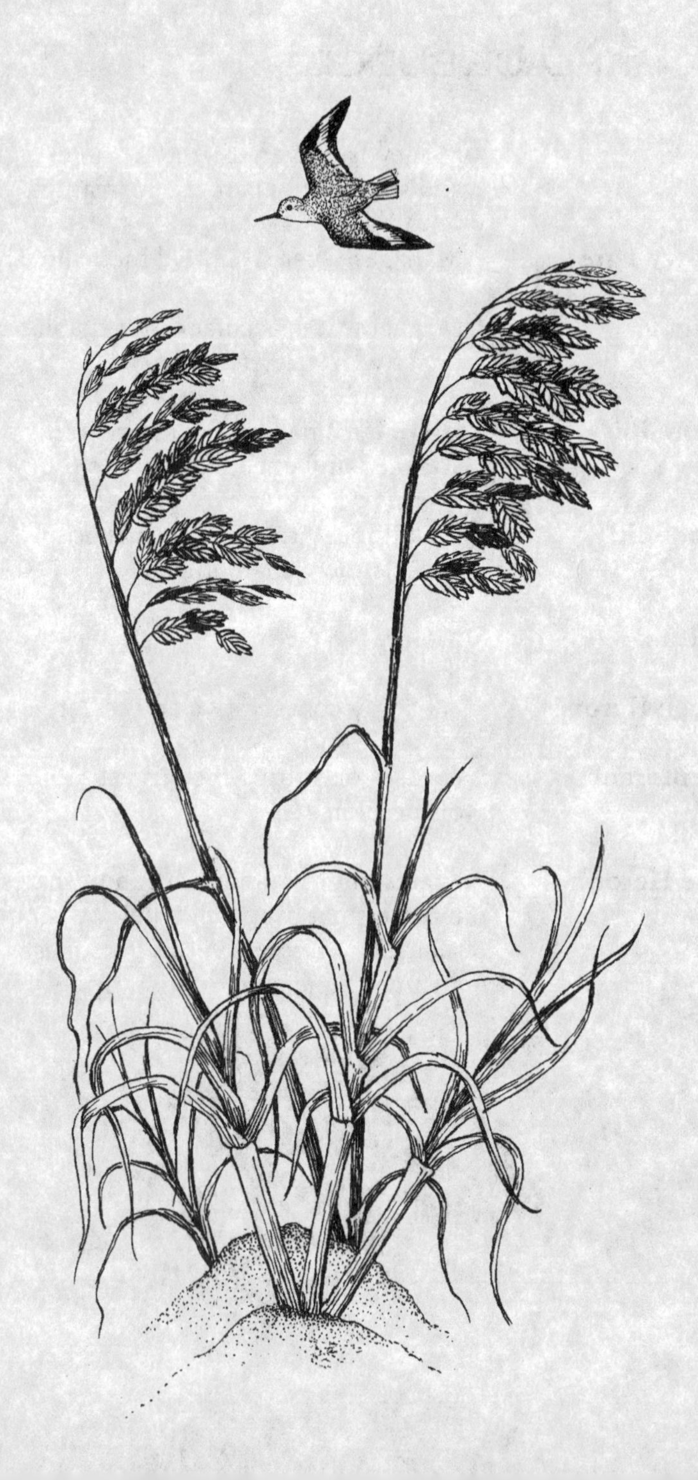

Matrix

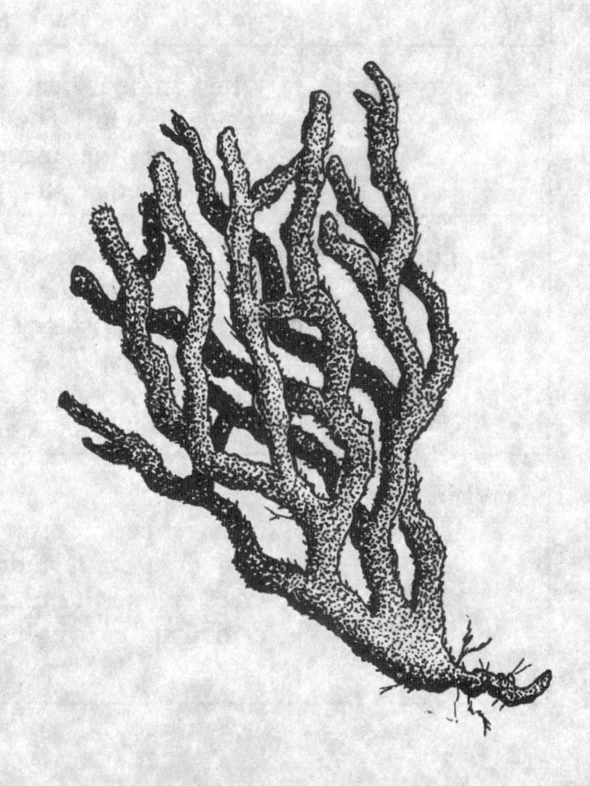

Interface Matrix

Each chart suggests the interface that will occur between the primary bird as supervisor and each of the other birds as subordinates. In each chart, there will be one occasion when the same behavior pattern applies to both the supervisor and the subordinate. The term "supervisor" is used generically. These charts will give you a quick look at style and strategies.

WOOD STORK SUPERVISOR

Behavior Patterns of Subordinates	Interface
Wood Stork	Power pair. Corporate Duo. Two strong egos; both like arguments and intimidation. Supervisor must win, so subordinate must present proposals in way that the senior can take ownership.
Snowy Egret	When unified, great pair; when angered, awful. The Egret will need his Stork's priorities made clear and the rationale for any compromises.
Seagull	The Seagull survives the periodic rages of a Stork manager with passive-aggressive resistance. Both tend to blame others. The Seagull should get the Stork to prioritize tasks. The Stork's praise would go a long way.
White Ibis	This is the usual combination of Stork manager and Ibis subordinate. It's an all or nothing combination; the Ibis either loves or hates his boss. The Ibis must be careful not to over-invest emotionally.

WOOD STORK SUPERVISOR

Behavior Patterns of Subordinates	Interface
Roseate Spoonbill	A Stork likes a Spoonbill's taste and emotional intensity; but they may fight a lot (or Spoonbill pout), because a Stork doesn't waste much time on people's feelings. The Spoonbill should produce on time.
Brown Pelican	A Stork manager can worry if Pelican improvises too much. A Pelican dislikes a Stork's micro-management, so the Pelican must earn his freedom by producing reliably.
Green Heron	Stork manager wants action, but Green Heron is slow with analysis & conclusions. Stork may publicly vilify the cautious Green Heron who blocks his hasty decision. Best strategy for Green Heron subordinate is to present information to Stork face-to-face privately. Regularly scheduled meetings could lessen the Stork's intrusiveness.
Cormorant	Stork manager can keep subordinate Cormorant in line, if he chooses to notice damage being done.
Great Blue Heron	Can be a healthy combination if Stork manager recognizes vision & contributions of GB Heron. GB Heron does not function well if exposed to tantrums, foul language, or crude jokes of a Stork.

SNOWY EGRET SUPERVISOR

Behavior Patterns of Subordinates	Interface
Wood Stork	These two black/white thinkers are prone to anger and fights, or to unified action. It behooves the Stork to listen carefully to assess what the Egret wants and do his homework.
Snowy Egret	These critical perfectionists are devoted professionals, but also put-down artists, so must avoid competing as to who is more perfect. They will be slow decision-makers.
Seagull	This is a combination of procrastinators: the Egret's concern with errors and the Seagull's ability to see many perspectives. An Egret manager's structure, plus going out of his way to praise, will make a good duo. The Seagull can shoulder some of the Egret's over-work because of a similarity in work style and effort.
White Ibis	The Ibis may feel the cold shoulder of the Egret who does not focus on social-feeling needs. The Ibis must ensure accurate, timely and productive efforts.
Roseate Spoonbill	This can be a dissatisfied, distressed duo because the Egret sees all of the Spoonbill's flaws and points them out; the Spoonbill resents Egret's cold approach. However, both may be energized by excellence.

SNOWY EGRET SUPERVISOR

Behavior Patterns of Subordinates	Interface
Brown Pelican	This is a combination of opposites: structure & spontaneity; solitude & networker. The Pelican would not be optimally productive with an Egret manager because he does not thrive in an atmosphere of criticism.
Green Heron	Although the Green Heron is put off by the Egret's focus on detail, he is pleased by the Egret's valuing of his expertise and advice. Together, they can produce a controlled decision, if the Egret can withstand the Heron's criticism.
Cormorant	The negative focus common to both may mutually irritate and boil over. The Egret loves rules and the Cormorant hates them. An Egret is happy to expose a Cormorant when rules are broken. Neither feels personal allegiance.
Great Blue Heron	An Egret supervisor likes a GB Heron's attention to detail, but worries that deadlines & minute detail may be missed due to the GB Heron's focus on performance. An Egret is uncomfortable in the public eye, so resents a GB Heron's ease. Care must be taken by the Heron to give credit publicly to his Egret.

SEAGULL SUPERVISOR

Behavior Patterns of Subordinates	Interface
Wood Stork	A Seagull manager can appearun focused or consensus-focused to a Stork, which he finds threatening. A Stork needs order and certainty. A Seagull generally dislikes supervising, so the Stork may take advantage (or try!).
Snowy Egret	A Seagull manager can chart the course and delegate tasks to an Egret for daily completion (which meets Seagull's needs and ensures that Egret doesn't overwork a project). An Egret has many skills and perspectives in common with a Seagull manager.
Seagull	Because Seagulls are rare as supervisors, the combination of boss and subordinate is unlikely. However, they can merge (or submerge) with their surroundings and, if sufficiently praised, will be very productive. This duo would tend to be collaborative rather than hierarchical.
White Ibis	A Seagull likes the warm support of an Ibis, so they can be a good pair, regardless of positions held. However, the Ibis can be frustrated by Seagull's slow decision-making.
Roseate Spoonbill	A Seagull manager would be hard for a Spoonbill to under- stand, because a Seagull is not focused on himself. A Seagull manager would be wise to give both structure and appreciation to his Spoonbill. The Spoonbill should ask for short-term mini-task deadlines to ensure having materials ready when the Seagull needs them for further coordination.

SEAGULL SUPERVISOR

Behavior Patterns of Subordinates	Interface
Brown Pelican	A Seagull manager likes routine & harmony. A Pelican may seem elusive to a Seagull, so he should check in regularly and complete tasks on time.
Green Heron	A Seagull likes systemized procedures, so a Green Heron can fare well with a Seagull as manager if he provides the information needed in a timely fashion. An additional benefit to the Heron is that he is relieved of most of the kind of details & reports he dislikes doing, but he must remember to be grateful.
Cormorant	A Seagull is the most likely profile to spot a Cormorant on the prowl and to take action if the Seagull perceives a threat to the workplace or to valued individuals. Because the Seagull supervisor likes routine & harmony, failure to provide accurate information in the required format on time is a cardinal sin.
Great Blue Heron	If a Seagull supervises a GB Heron, his slowness in decision-making will drive his Heron crazy. The Heron can help by laying out options in detail, thus relieving the Seagull of the burden of research.

WHITE IBIS SUPERVISOR

Behavior Patterns of Subordinates	Interface
Wood Stork	An Ibis manager of a Stork might occur in non-profit organizations where the founding mother/father still rules the roost. If the Stork can be less aggressive than usual, his can-do attitude may win the day. By doing operational duties, he can free his Ibis for the interpersonal work he does so well.
Snowy Egret	An Ibis manager may tend to focus on favorites; the Egret will resent having to earn favor above & beyond meticulous work. The Egret should get definitive guidelines and deadlines from his Ibis, and have a regularly scheduled time to meet weekly, so that the Ibis stays up on his progress.
Seagull	An Ibis manager could be impatient with the slow pace of a Seagull subordinate, but Seagulls are loyal & thrive on praise. The Seagull should have a scheduled daily meeting in order to know his supervisor's upcoming requirements.
White Ibis	This would be an unusual combination because both want to be supporters and have dependents. This could be a stalemate or a very warm relationship.

WHITE IBIS SUPERVISOR

Behavior Patterns of Subordinates	Interface
Roseate Spoonbill	The Ibis manager would be wise to give the Spoonbill specific areas of work for which she is responsible. These two have similar styles and paces of processing work, so meeting for short-term deadlines is the best strategy for peace and productivity.
Brown Pelican	These two are enthusiastic enjoyers, but an Ibis manager would need to ensure that a Pelican subordinate has sufficient structure to complete a task and that he does not get bored.
Green Heron	This combination works well with either being the manager or supervisor. One caution is the requirement for the Ibis to respect the Green Heron's need for privacy & space. The Ibis tends to be more intrusive by nature than is comfortable for the Green Heron.
Cormorant	The Ibis is at risk for manipulation by the Cormorant. Because the Ibis wants to be supportive, the wily Cormorant may take this innocent for a ride, getting the Ibis to do his dirty work, as he complains about his treatment by others.
Great Blue Heron	The Ibis manager may have a tendency to be intrusive on the GB Heron's privacy and space. If the Ibis will give the GB Heron a specific area of responsibility, so that he can achieve to his heart's delight, this will be a good combination.

ROSEATE SPOONBILL SUPERVISOR

Behavior Patterns of Subordinates	Interface
Wood Stork	A Spoonbill will feel either supported or threatened by the great drive and need for completion that a Stork brings to the duo. Both are spontaneous actors, but the Spoonbill may just change focus, not feeling compelled to complete the first passionate enterprise they started. A Stork will tend to see the Spoonbill as a walking soap opera, and while intrigued at first, will tire very quickly, especially if he is the brunt of blame, a common Spoonbill tactic when stressed. A Stork may need to leave before his reputation is damaged by his volatile Spoonbill.
Snowy Egret	The Spoonbill does permit an Egret to feed with his flock periodically, but does not allow him to be feisty. A wise Spoonbill supervisor will task the Egret with the details of joint projects because the Spoonbill dislikes them and things tend to fall through the cracks. The Egret can be a superb administrator for the creative supervisor.
Seagull	A Seagull will need consistent guidelines from his supervisor in order to deal with the Spoonbill's frequent creative changes in direction. A Seagull will need strategies for balancing the emotional intensity of his Spoonbill.
White Ibis	The Spoonbill will find the Ibis a good supportive worker, especially if the Ibis can be helpful on a personal level to the Spoonbill. The Ibis likes to give attention and the Spoonbill likes to receive it. Both need structure to be productive.

ROSEATE SPOONBILL SUPERVISOR

Behavior Patterns of Subordinates	Interface
Roseate Spoonbill	A Spoonbill pair would be extremely rare, except in aesthetic and artistic endeavors, where their mutual achievement could be quite remarkable. Usually, the Spoonbill's extreme intensity requires balance by those who are more structured and less self-focused.
Brown Pelican	This is a fairly common pairing; both are spontaneous and fun-loving and are drawn to aesthetic pleasures. However, the supervisor will have to provide some structure for their projects, so that their mutual creativity can be productive instead of just dreams.
Green Heron	A Green Heron will insulate and isolate himself as much as possible from a Spoonbill supervisor. They are opposites, one intellectual and one emotional. This pairing should be avoided if at all possible.
Cormorant	A Cormorant subordinate will have a heyday with a Spoonbill supervisor. He will be able to play on the supervisor's changeability and lack of attention, and on the Spoonbill's sense that he never gets his share. The Spoonbill is the most easily conned and manipulated by a Cormorant.
Great Blue Heron	The GB Heron will have to restrain himself from over-stepping boundaries of leadership because the Spoonbill's voids are so tempting. However, with the support of a GB Heron, a Spoonbill supervisor could excel with his great enthusiasm and aesthetic sensibilities.

BROWN PELICAN SUPERVISOR

Behavior Patterns of Subordinates	Interface
Wood Stork	The Stork can provide project support for his Pelican supervisor, whom he might perceive as unfocused. If the Stork can appreciate the different talents and style of his supervisor and bond for mutual benefit, this can be an effective pair....especially if the Pelican permits the Stork to "run the store."
Snowy Egret	The Pelican must provide sufficient structure for the Egret to know her marching orders (although the Pelican's style is more collegial than the Egret would prefer). The Egret can provide a superb administrative or research foundation for the Pelican supervisor so that his energies can be spent where his talents lie. If the Pelican will nudge the Egret's growth in domains where he fears imperfection, this can be an expanding experience for the usually fearful Egret.
Seagull	Much like the Egret, the Seagull needs specific tasks from the Pelican supervisor, as well as well-earned praise. This pair is common in entrepreneurial settings, because the Seagull anchors the operation and provides the follow-through, while the Pelican generates and markets the ideas.
White Ibis	Unless the Pelican supervisor provides structure for the Ibis, this pair tends not to be very productive, although they enjoy each other.

BROWN PELICAN SUPERVISOR

Behavior Patterns of Subordinates	Interface
Roseate Spoonbill	The Pelican supervisor often is an erratic day-to-day manager; if he delegates work to a Spoonbill subordinate, he must structure many small deadlines to ensure completion. They share the same strengths and weaknesses, so are not a good balance for each other, except that the Pelican is generally upbeat and the Spoonbill can thrive on his optimism.
Brown Pelican	Two Pelicans! What fun and what chaos!
Green Heron	The Pelican supervisor would be frustrated by the Green Heron's introversion and his need to consider all angles before taking action. However, the Pelican could be well-served by a Green Heron adviser before leaping to a decision.
Cormorant	The Pelican generally finds the Cormorant amusing and is usually not the target of a Cormorant. However, the Pelican supervisor does leave a lot of latitude for a Cormorant to be busy because the Pelican doesn't keep his eye on the store.
Great Blue Heron	This can be a high-performing duo, with the shared visions plus the structure that the GB Heron adds to bring projects to fruition.

GREEN HERON SUPERVISOR

Behavior Patterns of Subordinates	Interface
Wood Stork	The Green Heron supervisor relishes dispassionate analysis, and would be delighted to hand-off the operational follow-through to a Stork who is project-completion oriented. This pair can be exceedingly effective in the full continuum of research and development. However, the Stork subordinate will have to exercise patience, as the Green Heron won't hand off the execution portion until he is satisfied with his analysis (a slower process than when the two are reversed in roles).
Snowy Egret	The Egret subordinate may need more detailed work instructions than the cerebral Green Heron is inclined to give, so he will have to query his supervisor carefully. Both are introverts and do not waste much time on inter-personal communications.
Seagull	The Seagull will protect his Green Heron from inappropriate intrusions by others in the workplace and will see to all of the daily operations. The Seagull may need to make his own needs explicit for his Green Heron, who must go beyond his usual comfort zone to praise the support of his Seagull.
White Ibis	A White Ibis subordinate will feel unvalued by his Green Heron Supervisor, because the personal connectedness of the Ibis is not on the Green Heron's radar. The Ibis must clarify tasks given, as the Green Heron tends to give less concrete assignments than the Ibis needs. Not an easy pair.

GREEN HERON SUPERVISOR

Behavior Patterns of Subordinates	Interface
Roseate Spoonbill	Talk about opposites! The Spoonbill will loose his pink color (from lack of attention) and the Green Heron will isolate himself from the emotional intensity. Avoid this pairing.
Brown Pelican	A Green Heron supervisor will be frustrated by a Pelican's spontaneity, lack of respect for work schedules, and his seeming lack of focus. Not a healthy pairing for either bird.
Green Heron	Two Green Herons would honor each other's privacy, intellectual focus, and silence. They would be a good research pair. Someone else will be needed to see to their daily operations, however, as these two would be oblivious.
Cormorant	The Cormorant usually steers clear of a Green Heron because his motivations and activities are so readily apparent, should the Green Heron pause from his preferred intellectual pursuits long enough to notice. The Cormorant simply isn't in the Green Heron's league.
Great Blue Heron	This collaborative duo can be remarkably productive. The GB Heron's interpersonal skills can enable the Green Heron supervisor to "be all that he can be" without the threat of competition. The GB Heron honors the different knowledges, skills, and abilities of the Green Heron and seeks fruitful ways to bring those to fruition. If the Green Heron will likewise honor and utilize the GB Heron subordinate's talents, his visions can be marketed and executed.

CORMORANT SUPERVISOR

Behavior Patterns of Subordinates	Interface
Wood Stork	A Wood Stork subordinate will recognize a Cormorant supervisor quite readily. He will then need to assess whether he can totally disable his boss or not. If not assured of a win, then he must leave the workplace, as he will be damaged by a wounded Cormorant. A request for a lateral assignment is the usual technique, though a sabbatical or management training option is strategic.
Snowy Egret	A Snowy Egret subordinate will find a rule or regulation that will entrap the Cormorant supervisor and he will delight in it.
Seagull	A Seagull subordinate will notice the activities of the Cormorant and, where they impact people to whom he is loyal, the Seagull will target the Cormorant over and over until he successfully takes his fish.
White Ibis	A White Ibis subordinate will ponder awhile until he is sure that the Cormorant is behaving, as he perceives. Then, at all costs, he will protect those whom he supports.

CORMORANT SUPERVISOR

Behavior Patterns of Subordinates	Interface
Roseate Spoonbill	The Spoonbill will be oblivious to the Cormorant's activities. As long as the Cormorant provides the Spoonbill with special attention, he really doesn't care what is happening to anyone else.
Brown Pelican	The Pelican generally is not targeted by the Cormorant. However, if he finally notices what the Cormorant is doing, he may become an enraged fowl, a noteworthy opponent.
Green Heron	A Green Heron will tend to have insulated and isolated himself enough to ride out the tenure of a Cormorant supervisor. However, he has intellectual skills that should worry the Cormorant, should he become engaged.
Cormorant	These two deserve each other!
Great Blue Heron	Because the GB Heron tends to see people at their best, it takes awhile for him to deeply understand the games being played. Much like the Stork, he needs to assess if he can win with a lethal strike of his bill. If not, he should leave for his own well-being. He reminds a Cormorant, just by being who he is, what human beings at their best really can be and this enrages the Cormorant supervisor.

GREAT BLUE HERON SUPERVISOR

Behavior Patterns of Subordinates	Interface
Wood Stork	A Stork subordinate must gain more comfort with democratic processes and strategic planning than he prefers and careful inter-personal interface with others in the workplace. Arrogance will not be tolerated. This pair, if they can honor each other's strengths, can be a powerful team.
Snowy Egret	The GB Heron supervisor may give more global assignments than an Egret finds comfortable, so the Egret should query until he is sure what the Heron wants. If the Egret can tolerate some imperfection early in new tasks, he will find that his GB Heron can open new doors to success.
Seagull	The GB Heron is a supportive supervisor for his critical ally. He appreciates the efforts, effectiveness, and efficiencies that the Seagull has produced. The Seagull should ensure that they have scheduled times daily to keep each other informed.
White Ibis	The GB Heron supervisor understands the special "glue" of the Ibis. The Ibis also needs to demonstrate productive use of time, because his Heron is mission-focused.

GREAT BLUE HERON SUPERVISOR

Behavior Patterns of Subordinates	Interface
Roseate Spoonbill	The GB Heron supervisor should require regularly scheduled meetings with the Spoonbill, with tasks to be accomplished carefully delineated. The Spoonbill should ensure that he understands tasks assigned and when he needs help to meet deadlines that he confers with the Heron so he has a short-fused opportunity to correct any errors.
Brown Pelican	The GB Heron supervisor would be wise to provide the Pelican with structured deadlines, within which he is free to create results. The Pelican can add remarkable solutions to challenges, if given the freedom to perform on his own schedule. The Pelican, in order to earn this privilege, should be careful to communicate with the GB Heron about his progress between deadline meetings.
Green Heron	The GB Heron values the intellect of the Green Heron and will try to provide the insulation he needs to be productive. The Green Heron would be wise to keep his GB Heron apprised of his progress informally, rather than waiting for deadlines, so these two can brainstorm and move ahead as knowledge and visions intersect.

GREAT BLUE HERON SUPERVISOR

Behavior Patterns of Subordinates	Interface
Cormorant awhile for the the GB be number of the GB	The GB Heron supervisor may take to be sure of what his Cormorant subordinate is up to. It is tempting Cormorant to believe that because Heron's focus is elsewhere, he will oblivious to the Cormorant's machinations. However, any the birds noted above will inform Heron and his bill is a lethal tool.
Great Blue Heron	Generally, there are not two GB Herons in one workplace. As in nature, when a subordinate GB Heron comes upon a senior, he will respectfully withdraw, unless food is plentiful (mission challenges). They are more likely to encounter each other in gatherings from a variety of workplaces, such as conferences, symposia, and networks. In these settings, one sees respect, mentoring, and eager sharing of resources.

Cymbals & Symbols

Postlogue: Cymbals and Symbols

Birds of a feather flock together.

Miguel de Cervantes, *Don Quijote* (1605 –1615)

Birds of a feather will gather together.

Robert Burton, *Anatomy of Melancholy* (1621 – 1651)

Two of the great writers of the early seventeenth century noted the same phenomenon. Their observation is considered commonplace today, but it is only partially true. Birds of the same feather do recognize their commonality, but not all choose to be in close proximity. Some prefer to be solo. Others are communal in the rookery during nesting time only. Yet others are very congenial about sharing their terrain, as they know that their supply of food is unlimited or that what they seek is different from that desired by others.

Wherever and whenever I have talked about my research on this book, people have chuckled, had looks of recognition, and have said, "How soon will this be published? I need it now!"

Only at the end of the production cycle did I realize that my El Greco-esque elongated wooden statue of Don Quijote, purchased during my days in Spain (and an encouragement as I struggled to read Cervantes' *Don Quijote de la Mancha* in the original Spanish), looks like a Great Blue Heron with neck extended! In his search for truth and goodness, accompanied by folks who resemble our birds, he encountered the forces of ignorance, callousness, self-pity, poverty, petty tyranny, and evil – just as we all do. Cymbals and symbols are part of our religious rituals because their combination stirs us… just as Charley Harper comments on the back cover of this book: "Such analogies can be instructive, amusing, even frightening…"

Connections… be open to them because they will feed your metaphorical treasure trove – that unconscious network that will enrich, underscore, and enable fruitful action in your own endeavors.

Birding will never be the same!

Selected Bibliography

Although the author's observations of people and fowl are her primary source, hundreds of books and articles contributed to her insights. For further reading, the following are recommended:

Allen, Hayward. *The Great Blue Heron.* Minnetonka, MN: NorthWord Press, 1999.

Austin, Nancy K. "The Enneagram: Management's New Numbers Game." *Working Woman* (November 1995), 16, 21.

Baron, Renee and Wagele, Elizabeth. *The Enneagram Made Easy: Discover the 9 Types of People.* New York: HarperCollins Publishers, 1994.

Belenky, Mary Field, et al. *Women's Ways of Knowing: The Development of Self, Voice, and Mind.* New York: Basic Books, Inc., 1986.

Bolen, Jean Shinoda, M.D. *Goddesses in Everywoman: A New Psychology of Women.* New York: Harper Colophon Books, 1984.

Buckingham, Marcus and Clifton, Donald O., Ph.D. *Now, Discover Your Strengths.* New York: The Free Press, 2001.

Drucker Foundation, The, Editors: Hesselbein, Frances, Goldsmith, Marshall, and Beckhard, Richard. *The Leader of the Future: New Visions, Strategies, and Practices for the Next Era* . San Francisco: Jossey-Bass Publishers, 1996.

Farrand, John, Jr. *National Audubon Society Field Guide to North American Birds: Eastern Region.* Revised Edition. New York: Alfred A. Knopf, 1995.

Gibson, Graeme, *The Bedside Book of Birds – An Avian Miscellany,* New York: Doubleday (2005).

Gingras, Pierre. *The Secret Lives of Birds.* Buffalo, NY: Firefly Books, Inc., 1997.

Hamilton, Edith. *Mythology.* New York: New American Library, 1957.

Kiersey, David, Ph.D. and Bates, Marilyn. *Please Understand Me: Character & Temperament Types.* Del Mar, CA: P;rometheus Nemesis Book Company, 1984.

Korem, Dan. *The Art of Profiling: Reading People Right the First Time.* Richardson, TX: International Focus Press, 1997.

Palmer, Helen. *The Enneagram in Love and Work: Understanding Your Intimate and Business Relationships.* San Francisco: HarperSanFrancisco, 1995.

Palmer, Helen. *The Enneagram: Understanding Yourself and the Others in Your Life.* San Francisco: HarperSanFrancisco, 1988.

Parrott, Les, III, Ph.D. *High-Maintenance Relationships: How to Handle Impossible People.* Wheaton, IL: Tyndale House Publishers, 1996.

Parrott, Les, III, Ph.D. *The Control Freak: Coping With Those Around You. Taming the One Within.* Wheaton, IL: Tyndale House Publishers, 2000.

Peterson, Roger Tory. *A Field Guide to the Birds of Eastern and Central North America.* Boston: Houghton Mifflin Company, 1980. (Peterson Field Guides)

Riso, Don Richard. *Understanding the Enneagram: The Practical Guide to Personality Types.* Boston: Houghton Mifflin Company, 1990.

Ritberger, Carol, Ph.D. *What Color is Your Personality?* Carlsbad, CA: Hay House, Inc., 1999.

Sibley, David Allen. *The Sibley Guide to Bird Life & Behavior.* New York: Alfred A. Knopf (2001). (National Audubon Society)

About the Author:
Kathleen Parker O'Beirne

What are the ingredients for thinking and writing about strategic relationships?

Childhood explorer of creeks and beaches and keen observer of birds and plants.

Irreverent observer of people since her pre-teen years, much to her mother's dismay.

Army daughter, wife of a Naval submariner, and mother of two cherished children – lots of new people at each new assignment and lots of time alone to hone interpersonal and intrapersonal skills.

Her career development has included educational, civic, religious, entrepreneurial, and bureaucratic settings, with roles as an employee, volunteer, and self-employed writer, educator, & consultant.

Smith College (B.A.) and Wesleyan University (M.A.L.S.)

First Class Girl Scout and Scout leader for 7 troop years.

Associate Editor, *Family Magazine* for 12 years.

Public Affairs Officer and Program Manager (for spouse employment and volunteer management), Department of Defense Office of Family Policy and Support and Public Affairs Officer, Naval Underwater Systems Center, New London, CT.

Director, Navy Family Service Center, Naval Submarine Base, New London, CT.

Teacher/Instructor: junior high through university post-graduate levels (including the National Defense University).

Boards of Directors: USO World Board, Alumnae Association of Smith College, Southeastern CT Women's Network, Military Child Education Coalition, & Community Coalition for Children.

Leadership Awards: Navy Wife of the Year, Outstanding Woman of the Year (both Camden County, GA BPW and S.E. CT Women's Network), Department of the Navy Meritorious Civilian Service Award, Athena Award, and United Church of Christ CT "Recognized Woman."

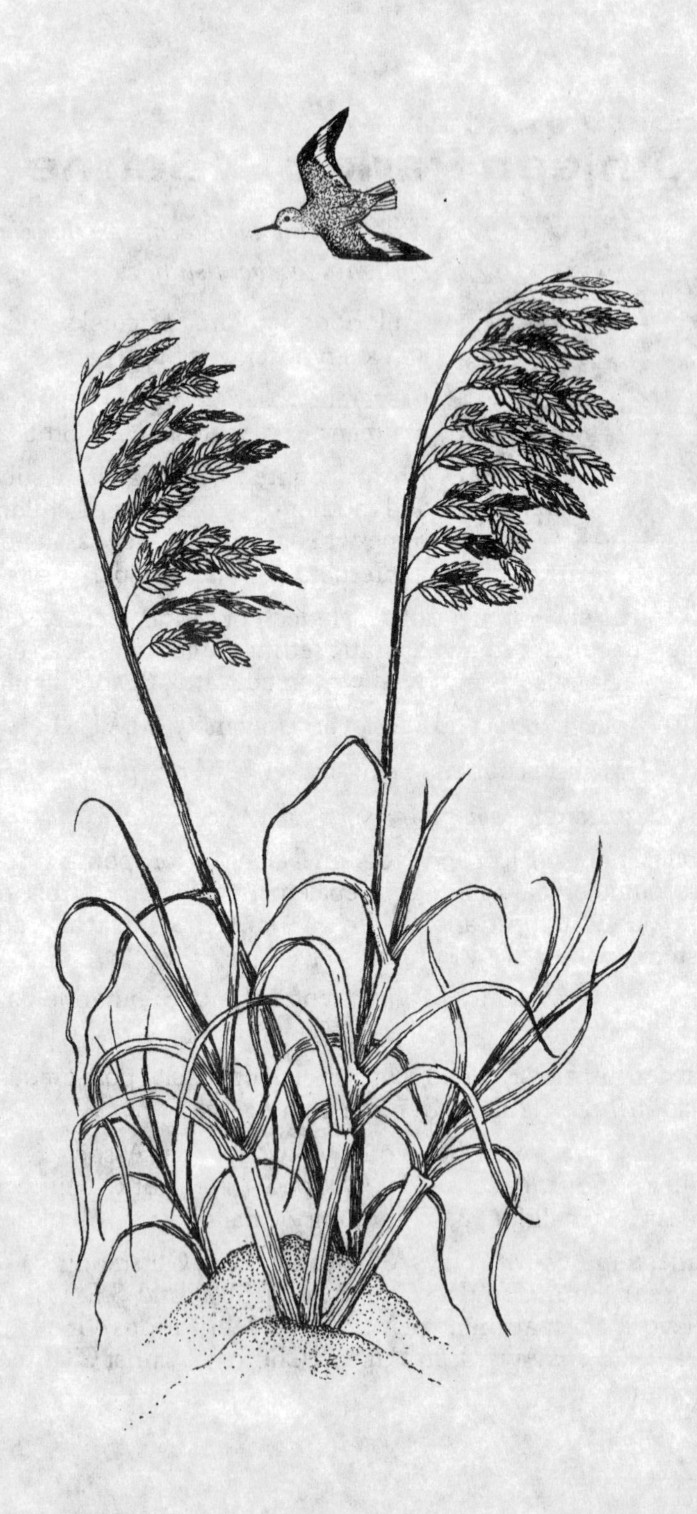

Re-order Information:

Life Is a Beach: Musings from the Sea

(ISBN 1-879979-09-8) $14.95

Birds of a Feather: Lessons from the Sea

(ISBN 1-879979-02-0) $19.95

Postage and handling per book: $ 2.00

Order by check or invoice to:

Lifescape Enterprises
P. O. Box 218
West Mystic, CT 06388

Queries to:

phone (860) 536-7179
fax (860) 536-2288
email: kathleenobeirne@aol.com

Credit card purchases:

To use your credit card, contact your local book store or

Bank Square Books Ltd (860) 536-3795
(49 W. Main St., Mystic, CT 06355)
Banksquarebks@msn.com

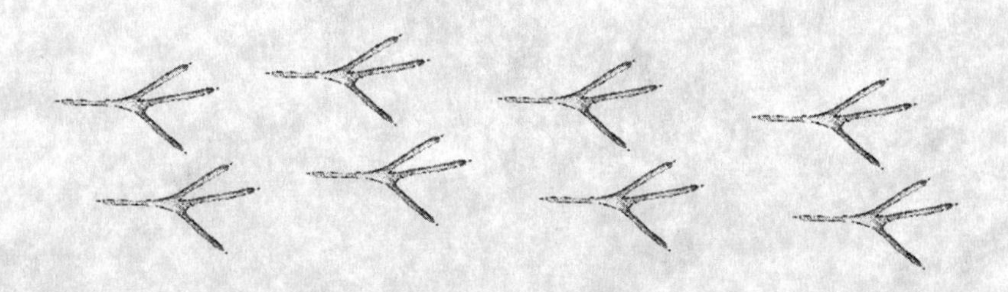

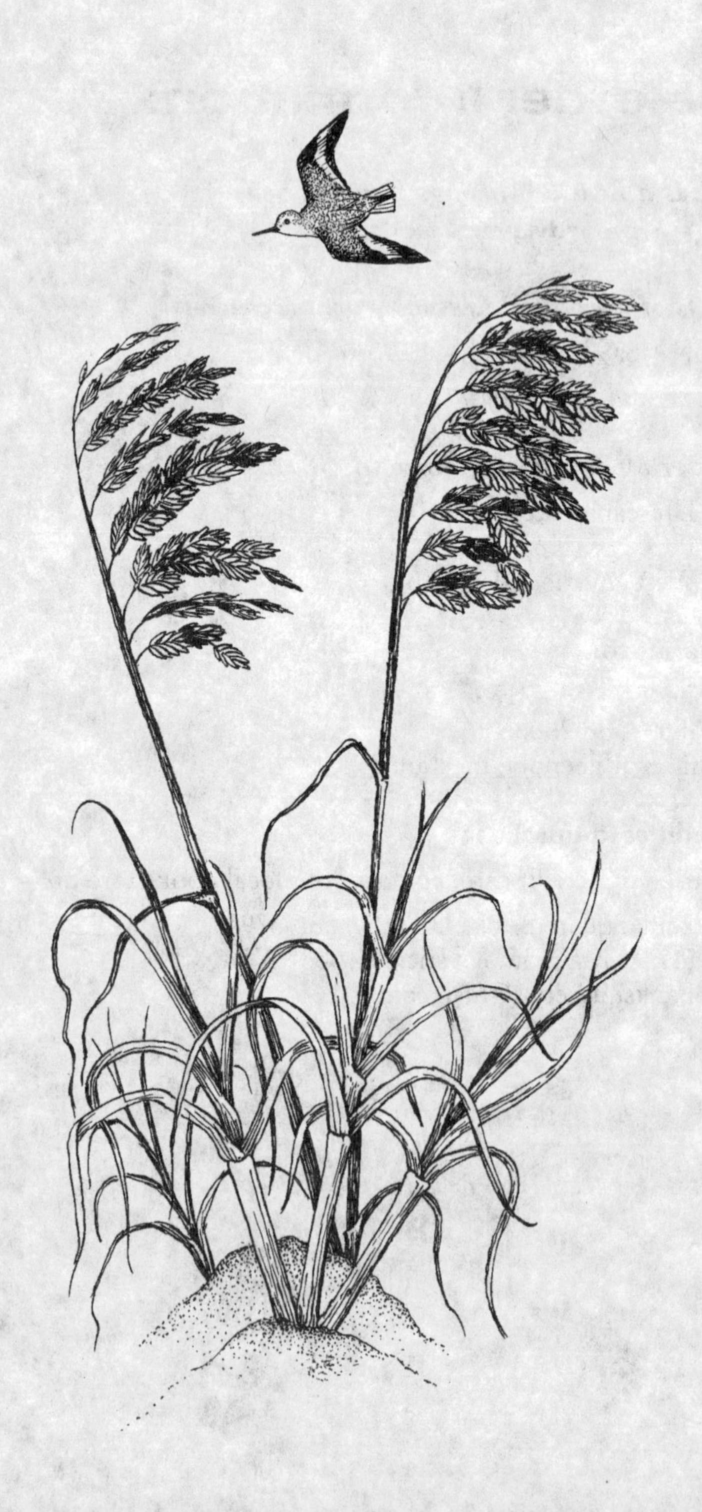

Re-order Information:

Life Is a Beach: Musings from the Sea
(ISBN 1-879979-09-8) $14.95

Birds of a Feather: Lessons from the Sea
(ISBN 1-879979-02-0) $19.95
Postage and handling per book: $ 2.00

Order by check or invoice to:

Lifescape Enterprises
P. O. Box 218
West Mystic, CT 06388

Queries to:

phone (860) 536-7179
fax (860) 536-2288
email: kathleenobeirne@aol.com

Credit card purchases:

To use your credit card, contact your local book store or

Bank Square Books Ltd (860) 536-3795
(49 W. Main St., Mystic, CT 06355)
Banksquarebks@msn.com

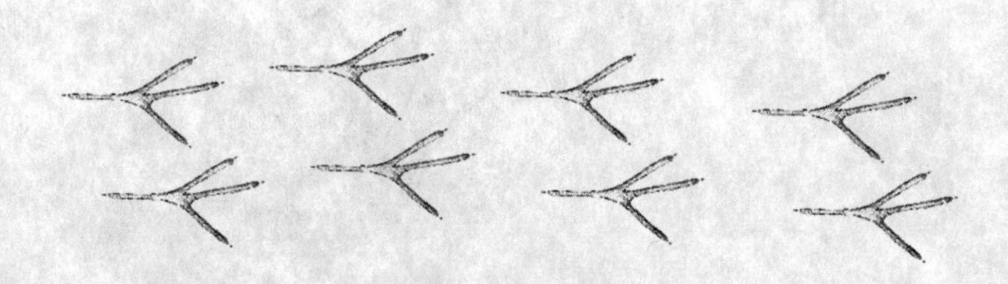

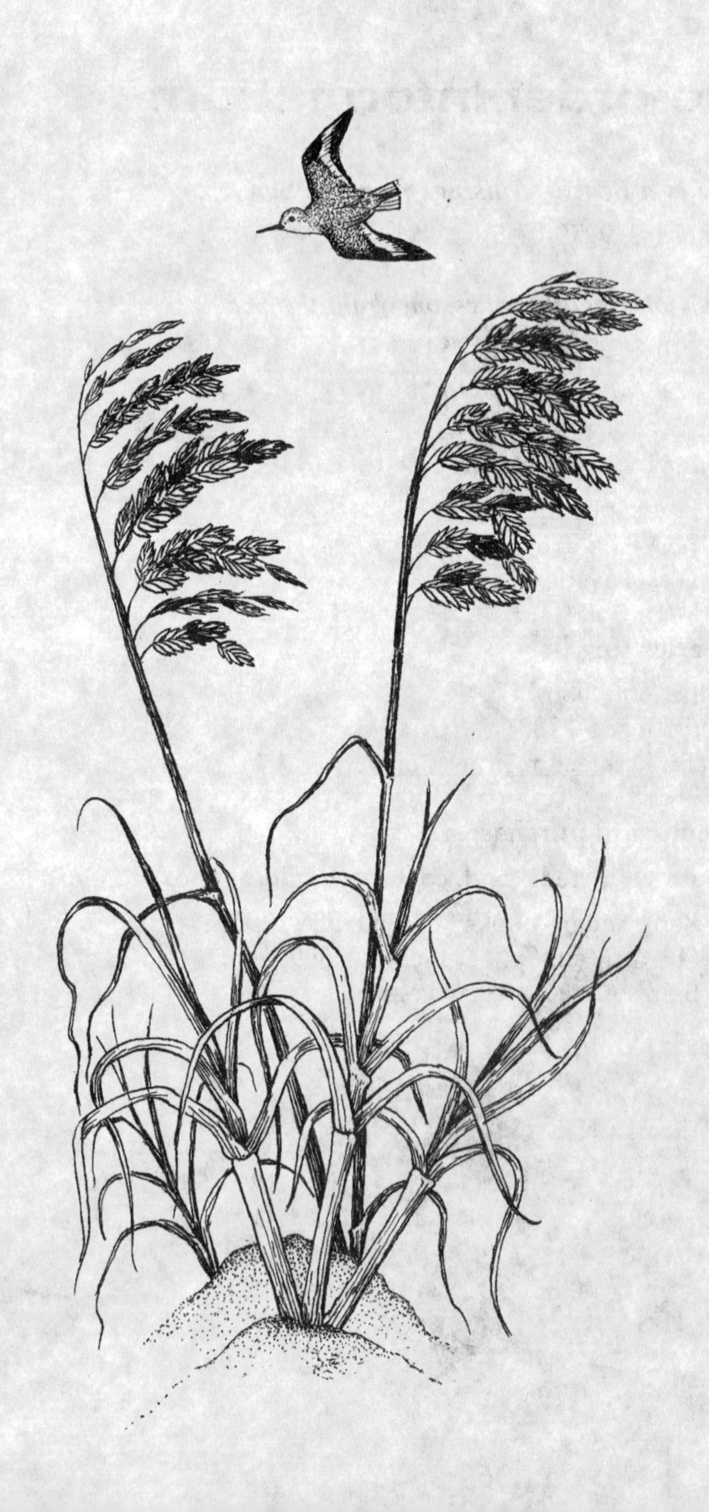

Re-order Information:

Life Is a Beach: Musings from the Sea

(ISBN 1-879979-09-8) $14.95

Birds of a Feather: Lessons from the Sea

(ISBN 1-879979-02-0) $19.95

Postage and handling per book: $ 2.00

Order by check or invoice to:

Lifescape Enterprises
P. O. Box 218
West Mystic, CT 06388

Queries to:

phone (860) 536-7179
fax (860) 536-2288
email: kathleenobeirne@aol.com

Credit card purchases:

To use your credit card, contact your local book store or

Bank Square Books Ltd (860) 536-3795
(49 W. Main St., Mystic, CT 06355)
Banksquarebks@msn.com

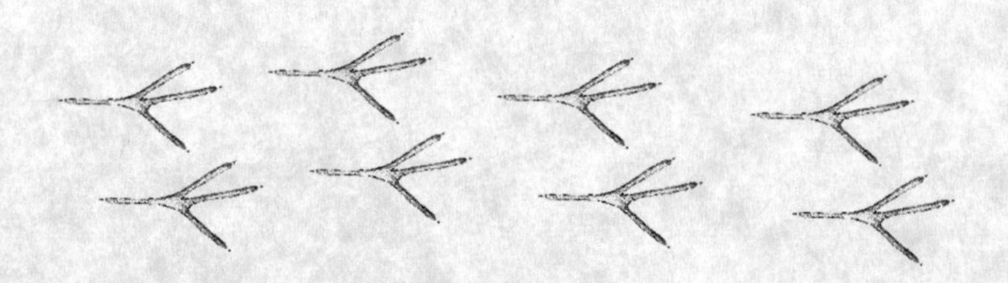

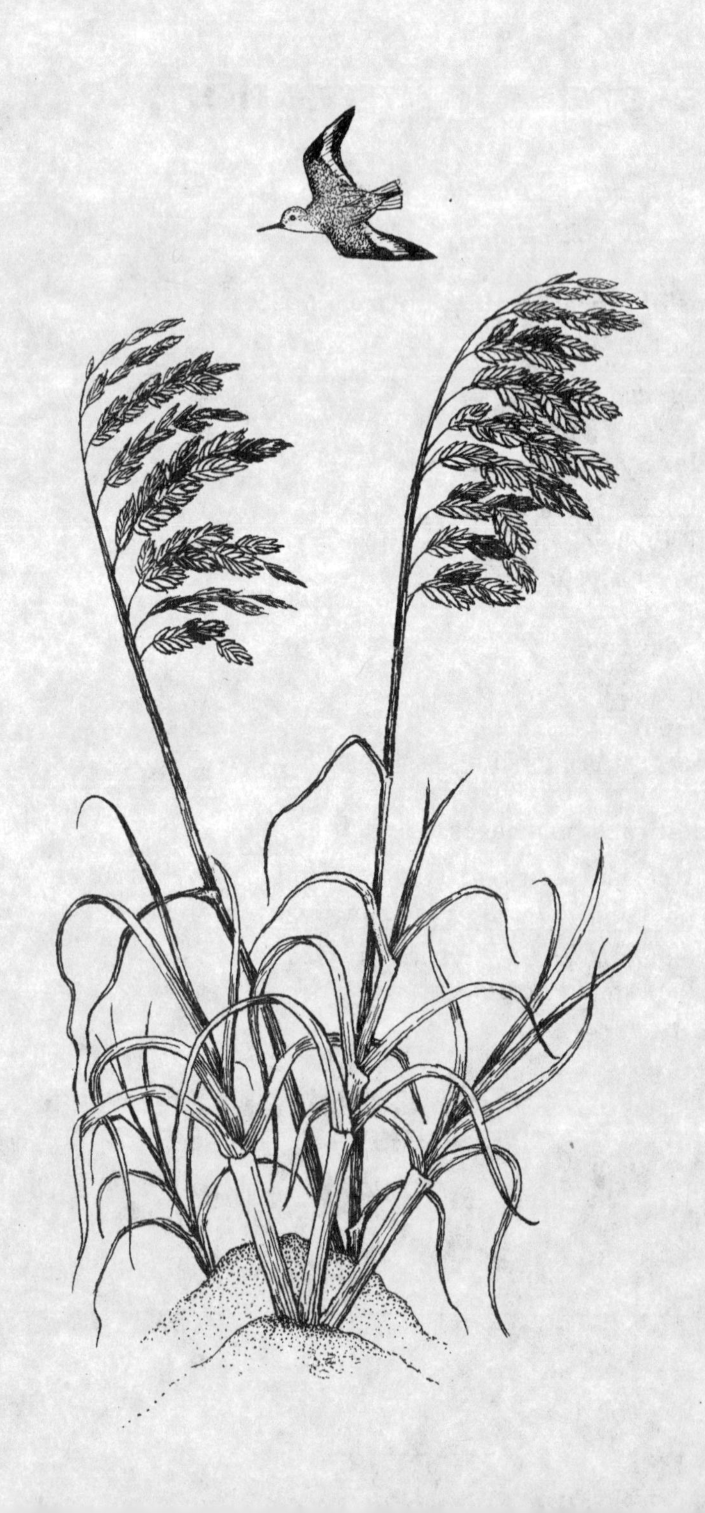